Machine Learning for Engineers

Ryan G. McClarren

Machine Learning for Engineers

Using Data to Solve Problems for Physical Systems

 Springer

Ryan G. McClarren
Department of Aerospace & Mechanical
Engineering
University of Notre Dame
Notre Dame, IN, USA

ISBN 978-3-030-70390-5   ISBN 978-3-030-70388-2   (eBook)
https://doi.org/10.1007/978-3-030-70388-2

This Springer imprint is published by the registered company Springer Nature Switzerland AG
The registered company address is: Gewerbestrasse 11, 6330 Cham, Switzerland

To my parents, Russell and Mary McClarren, for their love and support, as well as the supervised and unsupervised learning they provided me.

Preface

This book arose from the realization that students in traditional science and engineering[1] disciplines can both apply machine learning technology to problems that arise in their fields of study, and that the knowledge they have of science and engineering makes them well suited to understanding how machine learning works. The problem is that these students typically do not see these connections. Part of this is that most curricula do not include machine learning as a tool to solve problems alongside the mathematical methods and statistical techniques that are routinely used. The other part is that most courses and course materials that introduce machine learning motivate topics using examples that are foreign to students used to studying systems in the physical world. Many students would not find it immediately obvious that a model that can count the cats in an image might be useful for processing data from an experimental measurement.

This work attempts to meet science and engineering students in the middle by bringing machine learning techniques closer to problems in their disciplines and showing how they might be applied. The target audience is any student who has 1–2 years of instruction in a science or engineering discipline and is, therefore, familiar with calculus, a bit of differential equations, and a smattering of matrix algebra. There are points where more advanced topics are used, but these can be skipped by more junior readers. This book is also designed for more advanced students in other disciplines who want to see how machine learning might apply to their work. For example, graduate students (such as those in my research group) that are familiar with advanced computational science techniques for the solution of partial differential equations can use this book to learn how machine learning can augment their work.

---

[1] Here, by science and engineering I mean the science and engineering outside computer science and engineering.

Deciding on a scope for a work like this is a challenge. It is even more challenging in a field like machine learning where new results, models, and even new subfields appear with startling rapidity. The topics considered herein were chosen with the following two considerations in mind: (1) that the topics should give the reader enough background to confidently branch out into more complex versions of the topics discussed, and (2) that the topics have broad enough applicability to appeal to a large swath of science and engineering students and practitioners. This work does not present residual neural networks, but the discussion of convolutional neural networks that is covered is a good springboard for readers that have special interest in these models. There are several other examples of emerging trends (e.g., attention networks and physics-informed models using automatic differentiation) that are omitted from this work for the purpose of having a broad overview and appealing to the large set of potential readers.

The case studies and examples herein have the requisite data and code to reproduce them available on Github at https://github.com/DrRyanMc. The codes are in Python and make heavy use of Scikit-learn and Tensorflow, along with Numpy and Scipy. Any reader who is interested in plugging in their own problem into the models discussed in this work are encouraged to tinker with these examples.

The exercises at the end of each chapter come in two flavors. In the earlier chapters, there are problems that are used to demonstrate principles and facts regarding machine learning that most readers can show for themselves. There are also problems that are more like mini-projects suitable for a report-length deliverable. These mini projects often involve the production of new data and fitting a model. These problems are deliberately open ended and should be treated as reader-led case studies.

I want to acknowledge those that helped me in the production of this work. My students in the Notre Dame study abroad program in Rome during the spring of 2019 helped review a couple of early chapters, Todd Urbatsch gave me feedback as well, and Kelli Humbird has been gracious in answering several Tensorflow questions that I had. I also want to thank the Steinbuch Centre for Computing at Karlsruhe Institute of Technology, the Notre Dame Rome Global Gateway, and Universidad Adolfo Ibàñez for hosting me at different points during the creation of this work.

Finally, I am writing this preface while I am sequestered in my home during the events surrounding the COVID-19 pandemic during the early spring of 2020. I want to express my deep gratitude for those healthcare workers that are on the frontlines, as well as others up and down the global supply chain that make it possible for someone like me to work in relative comfort during such an extraordinary time.

Notre Dame, IN, USA   Ryan G. McClarren
April 2020

Contents

Acronyms

CIE	International Commission on Illumination
CNN	Convolutional Neural Network
DNN	Deep Neural Network
GRU	Gated Recurrent Unit
LSTM	Long Short-Term Memory
MAE	Mean-Absolute Error
MAPE	Mean-Absolute Percent Error
ML	Machine Learning
MNIST	Modified National Institute of Standards and Technology
MSE	Mean-Squared Error
OOB	Out-of-Bag
ReLU	Rectified Linear Unit
RL	Reinforcement Learning
RNN	Recurrent Neural Network
SGD	Stochastic Gradient Descent
SVD	Singular Value Decomposition
t-SNE	t-Distributed Stochastic Neighbor Embedding

Part I
Fundamentals

Chapter 1
The Landscape of Machine Learning: Supervised and Unsupervised Learning, Optimization, and Other Topics

He impaired his vision by holding the object too close. He might see, perhaps, one or two points with unusual clearness, but in so doing he, necessarily, lost sight of the matter as a whole.

—Edgar Allan Poe, "The Murders in the Rue Morgue"

Abstract In this chapter we survey the different types of machine learning problems and the way they are formulated. The two broad classes of *supervised* and *unsupervised* learning are covered in detail. For supervised learning we discuss loss functions, parsimony, and overfitting. The discussion includes regression problems for computing numerical output and classification problems in binary and multiclass forms. In the context of supervised learning we also discuss time series predictions and reinforcement learning. Unsupervised learning is presented as a way to find structure in data. The topics discussed include low-dimensional representations, embeddings, and association rules. To complement the discussion of learning problems there is a section on optimization as a means of fitting models as well as how to use machine learning models to optimize objectives. Sections on Bayesian statistics and cross-validation are included to aid in discussions later in the text. The discussion of cross-validation includes k-fold cross-validation, leave-one-out cross-validation, and how to apply cross-validation to time series as well as problems with unknown parameters in the loss function.

Keywords Supervised learning · Regression · Classification · Time series · Reinforcement learning · Unsupervised learning · Bayesian probability · Cross-validation

1.1 Supervised Learning

We begin with defining the types of problems that machine learning can solve. If we think of our data as a large box of toys and the machine learning algorithms as children, the problems we pose are the ways we present the toys to the children. In

© Springer Nature Switzerland AG 2021

R. G. McClarren, *Machine Learning for Engineers*,

https://doi.org/10.1007/978-3-030-70388-2_1

supervised learning, when the child grabs a toy, we ask what type of toy is it. When the child responds correctly, we say yes that is correct (and probably some other praise); if the child is incorrect, we say the correct answer. Eventually, the child will be able to develop internal rules for giving names to the toys. This rules will likely be more accurate on toys seen before but could also be used for extrapolation. If the child has learned the difference between "red blocks" and "blue blocks," it would be possible to properly identify a "red truck" when the child has only seen a "blue truck" before.

Analogous to the children's toy example, in supervised learning we have a set of data. Each data point, or case, has a number of features about it, called the independent variables that can be used to describe it. For toys it might be the shape, appearance, how you play with it, etc. For scientific data it may be properties of the system such as the material it is made out of, the mass, temperature, etc., as well as environmental factors, experimental conditions, simulation parameters, or anything else that varies from case to case.

In supervised learning the data must also have *labels*. These labels, also called dependent variables, can be a number, such as the measured values in an experiment or the result of a calculation or simulation, or they can be a categorical variable such as success versus failure, or membership in a group, i.e., is the leaf from a particular species of tree. These labels are the answers that we want machine learning to predict from the independent variables.

The goal of supervised learning is to have a machine learning model taking the independent variables as input and producing the correct labels for that case. We can measure the ability of the model to correctly predict the label using measures of accuracy, and we desire to have a model that has a high degree of accuracy, a point we will make more explicit later. A machine learning model has a number of parameters that need to be set to make a prediction. We use a set of data to set these parameters to maximize the accuracy of the model. The process of specifying the model parameters to fit the data is known as *training* the model. The data used to train the model is known as the training data.

Simply optimizing accuracy can lead to a problem known as overfitting. If we have too complicated a model, with a large number of parameters, the model can have a very high accuracy on the training data, but fail miserably when presented with new data. This can occur for a variety of reasons, but two primary causes are: (1) using a small training data set and having enough model parameters so that each point in the training data can be uniquely identified and (2) having training data that is not representative of the data that the model will see when we use it to make predictions. Consider the case where all of our training data has perfect measurements without noise, but when we want to use the model the independent variables will have measurement error and uncertainty. Given that the model has no "experience" with noisy inputs, the outputs from an overfit model in this case will be disappointing. The problem of overfitting is something we address in detail throughout our study.

In supervised learning we would also like to gain insight into our data. We are interested in questions such as:

- What are the relationships between independent and dependent variables?
- What independent variables is the prediction most sensitive to?
- Are there independent variables that do not affect the dependent variables?
- What additional training data would have the largest impact on improving the model?

The first of these questions is of keen interest to the scientists and engineers who want to be able to use the model to help understand how to affect the dependent variable. This could be in the form of a scaling law for an experiment or to help guide in the design of a physical system. The last question regards what experiments or data collection exercises we should perform to maximally improve our understanding of the relationship between the independent and dependent variables. This could be deciding which prototypes to build, small-scale experiments to field, or cases to measure in the field.

1.1.1 Regression

We now specify the supervised learning problem for dependent variables (labels) that are numerical values in some continuous range. Consider that we have collected the values of J independent variables, $x_1, x_2, \ldots, x_J$ and K dependent variables $y_1, y_2, \ldots, y_k$ at I cases. In that sense we have

$$\mathbf{x}_i = (x_{i1}, x_{i2}, \ldots, x_{iJ}), \qquad \mathbf{y}_i = (y_{i1}, y_{i2}, \ldots, y_{iK}), \qquad i = 1, 2, \ldots, I.$$
(1.1)

Our notation suggests that we can write this data set using matrices: $\mathbf{X}$ is a matrix of size $I \times J$ with each row being the independent variables for a case, and $\mathbf{Y}$ is a matrix of size $I \times K$ with each row being the dependent variables for a case.

Supervised learning attempts to find a function that maps from $\mathbf{x}$ to $\mathbf{y}$ as

$$\mathbf{y} = \mathbf{f}(\mathbf{x}).$$
(1.2)

The form of $\mathbf{f}(\mathbf{x})$ depends on the method used. Some methods prescribe a form for this function and as a result can only approximate certain classes of functions. In general the mapping will not be exact so we add an error term to Eq. (1.2):

$$\mathbf{y} = \mathbf{f}(\mathbf{x}) + \epsilon(\mathbf{x}, \mathbf{z}).$$
(1.3)

The function $\epsilon(\mathbf{x}, \mathbf{z})$ denotes the difference, i.e., error, between the function $\mathbf{f}$ produced by the supervised learning model and the true value of $\mathbf{y}$. This error could depend on the input variables, $\mathbf{x}$, but could also depend on variables that our model does not have access to that we denoted as $\mathbf{z}$. These unknown variables could have a large impact on the dependent variable, but they are not in our collected data set. As an example, consider a scenario where one wants to predict the temperature of an object that was outside. If we only had access to the ambient temperature, but did not record the amount of sunlight striking the object (making this an unknown variable to our model), we would have an error term that depended on the amount of direct sunlight.

The goal of the training procedure is to set parameters in the supervised learning model so that $\mathbf{f}(\mathbf{x})$ approximates the known values of $\mathbf{y}$ by minimizing a *loss function*. Throughout our study we will use L to denote a loss function. The loss function is typically a measure of the error, ϵ, but can include information about the complexity of the model, its sensitivity to changes, or its behavior in some limit. An example loss function is the squared-error loss function:

$$L_{se} = \sum_{i=1}^{I} \sum_{k=1}^{K} (y_{ik} - f_k(\mathbf{x}_i))^2 . \tag{1.4}$$

The squared-error loss function computes the square of the error for each dependent variable in each case and then sums up those errors over the cases. We use the square for two reasons: (1) we want to minimize the magnitude of the errors and (2) using a square gives smoothness to the loss function in terms of derivatives that some methods can take advantage of.

With a prescribed loss function, the training procedure then attempts to set the parameters in the supervised learning model to minimize the loss function over the training data. Consider that the model has P parameters, $w_1, w_2, \ldots, w_P$, that need to be set in the model. To minimize the loss function we seek to find sets of parameters, $\mathbf{w}$, where

$$\frac{\partial L}{\partial w_p} = 0, \qquad p = 1, 2, \ldots, P.$$

If the loss function is a convex function of the parameters, then we can be assured that the value of $\mathbf{w}$ is a minimum of the loss function. For non-convex optimization problems, there may be many local minima that are not the global minimum of the loss function. We will return to the topic of loss functions for each of the supervised learning models we study.

A potential loss function that includes information about the sensitivity of the prediction could take the form

$$L = \sum_{i=1}^{I} \left[\sum_{k=1}^{K} (y_{ik} - f_k(\mathbf{x}_i))^2 + \lambda \sum_{p=1}^{P} \left| \frac{\partial \mathbf{f}(\mathbf{x}_i)}{\partial w_p} \right| \right] . \tag{1.5}$$

This loss function balances squared error with the derivative of the model with respect to the parameters. The relative importance of the two terms is given using $\lambda > 0$. This loss function states a preference of the model builder: minimizing the error in the model is not the only consideration, and the sensitivity of the model to the parameters should also be minimized. This form of loss function attempts to create a model where the predictions will not change wildly when new data is introduced and the model is retrained. This can be an important consideration when a model is going to be used over time and retrained when new data is obtained. If the model produces very different predictions from each training period, it will be harder to believe the model's predictions.

Additionally, we can make the loss function penalize models that have too many non-zero parameters. The principle behind such a loss function is parsimony: the simplest model that can explain the data should be preferred to a more complicated model where possible. As in Occam's razor, the number of parameters should not be multiplied beyond necessity.[1] To have our loss function encourage parsimony we could add a term to the squared-error loss function:

$$L = \sum_{i=1}^{I}\sum_{k=1}^{K}(y_{ik} - f_k(\mathbf{x}_i))^2 + \lambda\sum_{p=1}^{P}|w_p|. \tag{1.6}$$

This loss function balances, through $\lambda > 0$, the error in the model with the magnitude of the parameters. Such a loss function can lead to trained models that have many parameters set to zero, and, in effect, a simpler model. In the next chapter we will see an example of this loss function in a type of regularized regression called *lasso*.

The principle of parsimony can also be applied to the number of independent variables that influence the model's prediction. If the independent variables that are available to the model are not all useful in predicting the dependent variables, the model should not include these variables when making a prediction. That is, changing these independent variables should not affect the model's value of $\mathbf{f}(\mathbf{x})$. In many cases, such situation corresponds with setting some of the model parameters to zero.

The squared error is not the only possible measure of the error. The measure of the error should always be positive, so that the error in one case does not cancel the error in another case, but that is one of the few hard constraints on the loss function. For example, we could take the absolute value of the error for each case, as opposed to the square. We can also include multiple considerations in the loss function by adding information about the sensitivity as well as encouraging parsimony.

There is one measure of the error that is often quoted for a regression model: R^2 or the coefficient of determination. This quantity is the fraction of the variance in the data explained by the model. The formula for R^2 is

---

[1] Occam's razor, commonly stated as *entia non sunt multiplicanda praeter necessitatem* (entities must not be multiplied beyond necessity), is named for William of Ockham, an English Franciscan Friar of the fourteenth century. However, the idea can be traced to much earlier statements, including one by Aristotle.

$$R^2 = 1 - \frac{\sum_{i=1}^{I} \sum_{k=1}^{K} (y_{ik} - f_k(\mathbf{x}_i))^2}{\sum_{i=1}^{I} \sum_{k=1}^{K} (y_{ik} - \bar{y}_k)^2}, \tag{1.7}$$

where $\bar{y}_k$ is the mean of the data for the kth output. A "perfect" model will have $R^2 = 1$ because the numerator in the second term will be zero. Also, the squared-error loss function attempts to make R^2 as close to one as possible because it is minimizing the numerator in Eq. (1.7). Nevertheless, a high R^2 does not necessarily mean that model is useful. The best advice is that a high value of R^2 indicates if the model might be useful, but further investigation is needed using the techniques we discuss below, such as cross-validation.

1.1.1.1 Overfitting

To see the need for parsimony we consider a simple relationship $y = 3x_1 + \delta$, where δ is random noise term given by a Gaussian distribution with mean zero. We have 50 realizations or cases, i.e., $I = 50$. For the independent variables we also have 50 variables so that $\mathbf{x} = (x_1, x_2, \ldots, x_{50})$, i.e., $J = 50$ where for $j > 1$ we have random values for x_j. If we fit a model on all 50 variables, we can get an "exact" fit so that $y = f(\mathbf{x})$ with no error, when in reality y only depends on x_1. In fact, we can fit a linear regression model (discussed in the next chapter) of the form

$$y = b + w_1 x_1 + w_2 x_2 + \cdots + w_{50} x_{50},$$

which will exactly reproduce the training data, even though there is no exact relationship between y and x because of the added noise. The reason we can get an exact model is that with 50 independent variables we have enough freedom in the model parameters to set them so each data point is reproduced.[2] This is shown in the left panel of Fig. 1.1. However, when we try to predict the value of y for new data, in the right panel of Fig. 1.1, the model gives completely unreasonable values. However, if we build a model with a loss function that seeks parsimony as well as accuracy, we get a model that is not "exact" for the training data, but much more accurate when presented with new data, $\mathbf{x}$, to predict y. This parsimonious model is a type of regularized regression called lasso regression that we will discuss in the next chapter. The end result of the parsimonious model is we have traded a bit of accuracy on the training data for robustness when applying the model to new data.

In this case the "exact" model would have a value of $R^2 = 1$ when looking at the training data. However, it is clear that this model is not perfect and that there are aspects to the model that R^2 does not capture.

---

[2] Technically, we only need 49 independent variables in this case because of the intercept, or bias, term in the model.

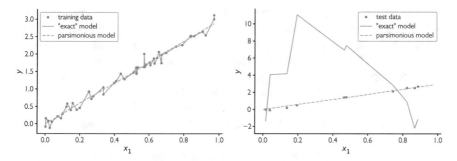

Fig. 1.1 Result of fitting a linear model with additional, unimportant, independent variables gives an "exact" match of the training data (left panel) but fails miserably when it tries to predict the relationship for 10 new cases not used to train the model (right panel). A parsimonious model has error in the training data but performs much better on the new, testing data

Regression
Regression is a supervised learning problem where we want to predict numeric value dependent variables given the values of independent variables. In these problems we define a loss function that specifies the magnitude of the error in the prediction as well as other considerations we find important (e.g., parsimony, low sensitivity to small changes in the variables).

1.1.2 Classification

In supervised learning problems for classification, we seek to assign a case to a group based on the independent variables. As before, we say that there are J independent variables, $x_1, \ldots, x_J$, and K dependent variables, $y_1, \ldots, y_K$. However, in classification the dependent variables can only take on a finite number of values. These values can be mapped onto the non-negative integers so that each dependent variable can take on a value $y_k \in \{0, 1, 2, \ldots, L\}$, where L is the maximum value y_k can take. An example of this mapping could be that y_1 was whether or not the system fails, $y_1 = 0$, or succeeds, $y_1 = 1$, in a given scenario described by $\mathbf{x}$. Or, if we wanted to predict whether an image defined by $\mathbf{x}$ has a cat, a dog, both, or neither, we could map this problem several ways: we could let $y_1 = 0$ be neither, $y_1 = 1$ be cat, $y_1 = 2$ be dog, and $y_1 = 3$ be both. Alternatively we could use two dependent variables and let $y_1 = 0$ for images with no cat, $y_1 = 1$ be for images with a cat, and let $y_2 = 0$ for images with no dog, and $y_2 = 1$ for images with a dog. Therefore, the condition of an image having neither a cat or a dog would be $y_1 = y_2 = 0$, and images having both would have $y_1 = y_2 = 1$.

Classification problems, though they are trained with dependent variables taking a finite number of values, will often produce the probability of the dependent variable taking *each* possible value. Then, if one wants the class which a case belongs, the highest value of the probability can be selected as the prediction for the class. Having the model produce probabilities is a feature that allows us to use the machinery of regression models discussed above. This is because we have transformed the problem of predicting an integer into one where we predict a probability that is a number between 0 and 1. Often, one more transformation is used to map the interval [0, 1] to the whole real line. Therefore, the model predicts a real number that is then transformed to a probability that is then used to assign the class.

To formulate a classification problem we need to specify a loss function. Because we are dealing with categorical variables, we need different loss functions than we used for regression. The most commonly used loss function for classification is the cross-entropy loss function. We first define the loss function for a binary classification problem with $K = 1$ and where the cases either have $y = 0$ or $y = 1$. We write predicted probability for case i that $y = 1$ as $h(\mathbf{x}_i)$; this implies that $1 - h(\mathbf{x}_i)$ is the probability that $y = 0$. The true value for case i is either $y_i = 0$ or 1. Therefore, the product

$$h(\mathbf{x}_i)^{y_i} (1 - h(\mathbf{x}_i))^{1-y_i}$$

would be 1 for a perfect model where $h(\mathbf{x}_i) = 1$ if $y_i = 1$ and $h(\mathbf{x}_i) = 0$ if $y_1 = 0$; this product will be less than 1 if the model is not perfect. We can get a measure for the error to minimize the model by taking the logarithm of this product and multiplying by -1. We then sum this result over all of the cases to get the cross-entropy loss function:

$$L_{\text{CE}} = \sum_{i=1}^{I} -\left[y_i \log h(\mathbf{x}_i) + (1 - y_i) \log(1 - h(\mathbf{x}_i)) \right]. \tag{1.8}$$

In this case, the loss will be zero when the model is perfect and greater than zero otherwise.

For a multiclass classification problem where y can take on many values, we instead use the model to predict a real number $z_\ell(\mathbf{x})$ for $y = \ell$ for $\ell = 0, \ldots, L$. This real number is defined such that as $z_\ell \to \infty$, the probability that $y = \ell$ goes to 1, as $z_\ell \to -\infty$, the probability that $y = \ell$ goes to 0, and $z_\ell > z_{\ell'}$ implies $y = \ell$ is more likely than $y = \ell'$. We then pass these values z_ℓ through the softmax function to map them each to the interval [0, 1]:

$$\text{softmax}(\ell, z_0, z_2, \ldots, z_L) = \frac{\exp(z_\ell)}{\sum_{\ell'=0}^{L} \exp(z'_\ell)} \in [0, 1], \tag{1.9}$$

where $z_\ell = z_\ell(\mathbf{x})$ because the predicted value of z_ℓ depends on the independent variables. The softmax function also has the property that

$$\sum_{\ell=0}^{L} \text{softmax}(\ell, z_0, z_1, \ldots, z_L) = 1,$$

so that we can interpret the softmax of each z_ℓ as the probability that the case has $y_i = \ell$. Now for a given case, the loss from the prediction is

$$\text{softmax}(0, z_0, z_1, \ldots, z_L)^{\mathcal{I}(0, y_i)} \times \cdots \times \text{softmax}(L, z_0, z_2, \ldots, z_L)^{\mathcal{I}(L, y_i)}$$

$$= \prod_{\ell=0}^{L} \text{softmax}(\ell, z_0, z_1, \ldots, z_L)^{\mathcal{I}(\ell, y_i)}, \qquad (1.10)$$

where the indicator function is

$$\mathcal{I}(\ell, y_i) = \delta_{\ell, y_i} = \begin{cases} 1 & y_i = \ell \\ 0 & \text{otherwise} \end{cases}.$$

The product in Eq. (1.10) will be 1 once if the model gives a probability of 1 for the correct value and probability of 0 for every other value. As before we take the logarithm and multiply by -1 for each case and then sum over all cases to define the general cross-entropy loss function:

$$L_{CE} = -\sum_{i=1}^{I} \sum_{\ell=0}^{L} \mathcal{I}(\ell, y_i) \log \text{softmax}(\ell, z_0, \ldots, z_L). \qquad (1.11)$$

The cross-entropy loss function provides a way to measure the accuracy of the model, and a function to minimize to train the model. However, as with regression we may want to add terms to the loss function to promote parsimonious models or reduce sensitivity to parameters in the model. To accomplish this we can add the same terms to the loss function as we discussed above for regression.

Classification
Classification looks to build supervised learning models to predict membership in a certain class based on independent variables. These models typically produce probabilities of a case belonging to a class.

- The loss functions for classification are different than those for regression.
- Binary (i.e., problems with two classes) and multiclass classification problems have connected, but slightly different approaches to solution.

1.1.3   Time Series

Times series are sequences of data points collected in a particular order. They have the property that for each variable evaluated at time t, we may know the values for all times $t' < t$ but do not know the values in the future. Examples of time series could be the reading of a gauge during the operation of a machine, the images in a movie, and even the words that form a document. In a sense, modeling time series using machine learning is the same as supervised learning with regression or classification. The difference is that in a time series the current state and previous states of the series are necessary components in the prediction. For example, in predicting the reading of a temperature gauge, the current temperature in the system, as well as the recent changes in the temperature and other independent variables, would be necessary to include in the model.

Using the current state of the dependent variable in the model as an independent variable introduces some changes in the approach to machine learning. In practice, as we will see, machine learning for time series requires the model to determine how long of a memory to have during the training process. Some processes only depend on the current system state to predict the next state, but others require knowledge of several states in the past (i.e., it must have memory). This requires specific types of machine learning systems, and we will cover this in some detail in a later chapter.

1.1.4   Reinforcement Learning

Reinforcement learning is a method to train machine learning to accomplish a task without specifying the correctness of any single decision. Consider a machine learning system that is trying to teach a bipedal robot how to move between point a to point b. Given the current position of the robot, other information about the surroundings, and its goal, machine learning will need to give the robot the next movement to make. We could use the training process to reward each step if it takes the robot closer to point b, but the resulting machine learning model would likely make it so that the robot would have a hard time navigating around obstacles where it would have to backtrack to accomplish the goal. We can instead define a loss function that rewards *all* the steps in a trial where the robot successfully moved from point a to point b, and "punishes" all the decisions in the trials where the robot does not successfully navigate from point a to point b. In this case the machine learning model will be trained to make moves that will ultimately accomplish the goal.

The problem with reinforcement learning is that it can be very slow to train. In the robot example we would have to give the robot many thousands of training exercises, different points a and b, different obstacles, etc., to amass the data to train the machine learning model. Part of this is that we will need some amount of success to find the decisions that should be rewarded. Usually these machine learning models begin with making random decisions, so it may take many trials before the model gets a success. Indeed, many of the famous successes of reinforcement learning, including the playing of video games [1] and Go [2], are in application domains where many trials can be generated via computer. The machine learning can play video games continuously and with many machines playing in parallel to generate the training data, and Go machine learning models can play against each other. Therefore, it is possible to have millions of cases for the purposes of training. There are many more games that a human could play in a lifetime, and as a result the machines played Go differently than expert humans did. Our study of reinforcement learning will apply these principles to science and engineering problems.

1.2 Unsupervised Learning

In supervised learning we know what we want the machine learning to accomplish: we want it to give us the values of the dependent variables given the independent variables. We formulate a loss function to accomplish this task and then train the model to minimize that loss. Even if we do not know exactly how to accomplish the task, as in reinforcement learning, we have answers that the model is trying to reproduce. In unsupervised learning we do not have the answers.

Unsupervised learning requires us to define a goal, but we do not need to have answers. A common unsupervised learning task is for the machine learning algorithm to find structure in the data. This structure could be answers to questions such as which variables have linear or nonlinear relationships between each other, what are the natural clusters or groupings of the data, are there redundant independent variables, etc.

An admittedly abecedarian application of unsupervised learning could be the situation where we have a set of photographs in a person's smart phone. There is likely to be multiple pictures of a small number of people such as close friends and immediate family. An unsupervised learning task could be to find the people who appear in multiple images and group the images based on which of those people is in the image. In this task notice we do not have to identify the number or identity of the people, and we just set the goal. It is possible that there is no underlying structure in these pictures, but we want our unsupervised model to look for structure.

1.2.1 Finding Structure

To specify the goal of the unsupervised learning exercise we need to formulate a loss function. This will depend on the type of application because the loss function is the primary tool we have in unsupervised learning. For example if we want to look for structure in the data, we may search for a set of variables that will nearly replicate the original data set. Consider that the initial data set is given by I cases of J variables: $\mathbf{x}_i = (x_{1i}, \ldots, x_{Ji})$. The goal of unsupervised learning may be to learn two transformations, $\mathbf{f}$ and $\mathbf{g}$, that define the mapping to a new set of variables $\mathbf{z}$ given by $\mathbf{f}(\mathbf{x}) = \mathbf{z}$, and the mapping back to the original variables $\mathbf{g}(\mathbf{z}) = \hat{\mathbf{x}}$. In this example, if the size of $\mathbf{z}$, J', is smaller than J, and the difference $\mathbf{x} - \hat{\mathbf{x}}$ is small, then we can consider that the unsupervised learning model found simpler, underlying structure in the data. To express this task in a loss function we could seek to minimize the least squares error in the reconstruction

$$L = \sum_{i=1}^{I} \sum_{j=1}^{J} (g_i(\mathbf{f}(\mathbf{x}_i)) - \mathbf{x}_i)^2 + \lambda J', \qquad \lambda > 0. \tag{1.12}$$

This loss function attempts to minimize the difference between $\hat{\mathbf{x}}$ and the original data $\mathbf{x}$ while also considering that the size J' should be minimized. If we did not include the $\lambda J'$ term, the loss function would be zero if the functions $\mathbf{g}$ and $\mathbf{f}$ were just multiplication by an identity matrix. Picking the value of λ gives a means to balance the reconstruction error with the size of the transformed variables $\mathbf{z}$.

If we do find functions $\mathbf{f}$ and $\mathbf{g}$ where the reconstruction error is small, we can consider $\mathbf{z}$ to be a low-dimensional representation of our original data. That is we can just store $\mathbf{z}$ for each case and use $\mathbf{g}(\mathbf{z})$ when we need the original values $\mathbf{x}$. This can require much less storage of data when $J' \ll J$ and can also improve performance on supervised learning problems using $\mathbf{z}$ as the independent variables rather than $\mathbf{x}$ because of the smaller dimensionality.

Another example of finding structure in data is the concept of embedding. Embedding refers to a type of model where large vectors of discrete numbers (e.g., integers) are mapped to a smaller vector of real numbers. For illustration, we could take a data set consisting of vectors of length 1000 containing only zeros or ones and map it into a two-dimensional vector of real numbers. The embedding is also specified so that the vectors of real numbers have large dot products when the original vectors are "similar." The definition of similar depends on the type of vectors being embedded. Similarity may be the dot product of vectors in the high-dimensional space, but it may also include other information such as relationships between the vectors measured in another way. One approach to embedding, t-SNE (t-distributed stochastic neighbor embedding), embeds high-dimensional vectors into a 2-D or 3-D space so that data can be visualized. This embedding has the effect that vectors in the high-dimensional space are close to each other in the 2-D or 3-D space; the net result is that natural clusters in the data can be visualized.

1.2.2   Association Rules

Another type of unsupervised learning task is finding association rules. This type of problem asks if there are particular combinations of variables that appear together, or are associated with each other. A simple illustration of this is the market basket problem. Here we consider the data generated by taking the inventory of the purchases of each person at a grocery store over the course of the day. We construct a data matrix of size $I \times J$ where I is the number of unique shoppers and J is the number of items in the store. Each row corresponds to a shopper, and each column in the data set represents an item (e.g., milk, eggs, bread, broccoli, etc.) with the number in row i, column j corresponding to the number of items j shopper i purchased. In this case we want to find items that are frequently purchased together and estimate the percentage of time these items go together, so that we could optimize the store layout, suggest purchases to customers, and perform other tasks. An example association rule could find that 40% of the time when someone purchased milk and eggs, they also purchased bread. We could use this information to place these three items together in the store, for example. Online stores and video streaming services use similar technologies to suggest items to you based on the purchases of others or the watching habits of others.

Association rules are also at work beyond many natural language processing tasks. Much of language construction is an association rule. Sentences have a finite number of structures and some words naturally follow others. Using unsupervised learning we can feed in many examples of text and it can learn how language is structured. This can be illustrated if we consider a data set that has a row for each sentence in a collection of texts (often called a corpus). The columns have an integer that corresponds to the words used in each position of the sentence in that the first column has an ID for the first word in the sentence, the second column has an ID for the second word in the sentence, and so on. There is also usually a special integer to identify the end of the sentence.

The association rule we are after, in such a case, will give probabilities for the next word in a sentence, given the current words in a sentence. These association rules can be used to write sentences. For example, we have an "association rule" for the first word of the sentence (basically a probability for each word in our vocabulary being the first word in a sentence). We randomly sample a starting word based on these probabilities. Then we have an association rule that tells us given that the first word what are the probabilities of the second word being a particular word. We can then sample from these probabilities a second word and continue using association rules until we sample the end of sentence word. This description is not meant to be a complete description of how natural language systems work in practice; there has been too much research to do those systems justice in our brief coverage here, but rather to use language as an example association rule.

In science and engineering we may want to learn association rules to understand a complex system's behavior. One case may be a manufacturing process where we are interested in understanding the differences among operators of machinery in the process. We can learn the ways that the operators behave using association rules. Perhaps there are conditions that lead to some operators controlling the system in a certain manner, while other operators behave differently. We could build a system of association rules that uncovers a series of rules such as "when conditions A and B occur, take action C" and so on. The analogy could extend into the ways that users operate an engineering design code. Certain users with long experience probably have a set of rules they follow when setting up a simulation in terms of parameters, meshing, tolerances, etc. that they may not be able to express in words. Using association rules, it may be possible to understand how these experts practice their craft.

With association rules the problem of scale can quickly become apparent. If I have J different variables that one wants to look for association rules between, there are $J!/(2(J-2)!)$ two way association rules to check, and $J!/(3(J-3)!)$ three way association rules, and so on.[3] For a modest-sized data set of $J = 100$, this leads to 4950 two way rules and 323,400 three way rules. Clearly, good algorithms are needed to find the plausible rules by quickly eliminating those rules where there are no examples in the data set.

In the context of association rules vector embeddings can also play a large role. Rather than representing a word as a vector with length equal to the number of words in the vocabulary, we can find embeddings that have large dot products when words appear nearby in the text. The representation as a vector of real numbers with this property gives the representations meaning that a vector of zeros or ones does not have. The embedding can also work for the market basket problem to reduce the data from potentially huge vectors of integers to smaller dimensional real vectors.

Unsupervised Learning
Unsupervised learning looks for structure in the data. These machine learning problems do not require an independent/dependent variable relationship (i.e., no labeling of cases is required). Unsupervised learning can be used to find:

- Low-dimensional representations of the data (i.e., transformations to the variables that capture the variability in the data)
- Vectors of real numbers called embeddings to represent large, discrete valued variables
- Association rules to know when certain conditions imply another condition to be more likely

1.3   Optimization and Machine Learning vs. Simulation

The concept of optimization is important in machine learning, and we have already hinted at where it is needed. Once the loss function is specified we need an optimization algorithm to find the parameters in the model that minimize the loss function. This is a difficult task because the optimization problems in machine

---

[3]These counts assume that A implies B is the same as B implies A because we have no basis to assume a causal relationship, and we are really just looking for correlations.

learning are often non-convex, as we have mentioned before. A non-convex optimization problem may have many local minima that are far away in magnitude from the global minimum. This means that the optimization problem must not simply look for a minimum, rather it must be resilient against getting stuck at a local minimum. There are a variety of optimization techniques that have been developed for or adapted to machine learning that we will discuss throughout our study. We will generally be using these as tools to accomplish our goal, rather than methods that we wish to develop, but we must be cognizant of the problems of non-convex optimization.

We will also need optimization to apply the outputs of machine learning as well. If we are designing a system, and we have a machine learning model that predicts the system performance over a set of objectives, we may want to apply an optimization algorithm to the machine learning model to find the optimal point in design space for our system. This procedure is similar to the standard procedure of applying optimization to a design code or simulation, except the simulations are replaced by the machine learning predictions. In principle we should be able to explore more of design space using machine learning than the full simulation due to the fact that making predictions using machine learning should be faster than performing a complete simulation.

In using machine learning to optimize a system one could argue that the machine learning will not be exact and is missing some of the rich physics, chemistry, biology, or other domain expertise that was built into the full simulation. While it is possible to formulate machine learning problems to respect laws of nature such as conservation of energy, even if the machine learning does not have such a principle automatically satisfied, it can be a useful design tool. To see this, we consider a scenario where a supervised learning model is trained on simulation data to predict a set of outputs. Then using optimization on the supervised learning model we find several candidates for optimal designs. We then run these designs through the full simulation to check them. If the machine learning predictions are acceptable designs, we have saved much searching in design space. If the machine learning predictions were incorrect, then we can retrain (i.e., improve) the model and search again. This procedure will only require enough simulation data to train the model and then only incremental simulation runs are needed. Nevertheless, both tools are necessary in the scenario described.

The usefulness of machine learning is even greater if uncertainties in the inputs need to be considered. Because machine learning can evaluate different input cases very rapidly, we can perturb the inputs of a design and see how robust they are to slight changes in conditions. It is possible that the "optimal" design, when there is zero uncertainty in the inputs, is different than the optimal design when robustness to perturbations is required. Indeed, machine learning has been used to find novel approaches to designing fusion energy experiments when a robustness metric is included: under perfect conditions a perfectly spherical experiment gives the best energy production. However, if one considers the robustness to perturbations to the experiment, as well as the energy production, machine learning models were able to find a time-dependent geometry that gives nearly the same energy production, but is robust to perturbations [3].

Optimization

Optimization plays a key role in machine learning:

- The training process is optimization: we try to minimize the loss function.
- Machine learning models can be used to find optimum values of the functions they are trained to learn.

1.4 Bayesian Probability

One concept that is important and widely used in machine learning methods is Bayesian probability. The idea of Bayesian probability is that we begin with an estimate of the behavior of the probability for a random variable and update our estimates once we have collected more data. We can consider the parameters in a machine learning model to be random variables because we have no expectation that there is a "correct" value for these parameters and that these parameters exist as distributions. Consider a parameter w that we consider to be between -1 and 1, but we have no other information about this parameter. We specify our knowledge of this parameter through a *prior* distribution. In this case we may specify the prior distribution as the probability density function

$$\pi(w) = \begin{cases} \frac{1}{2} & -1 \leq w \leq 1 \\ 0 & \text{otherwise} \end{cases}.$$

This function is a proper probability distribution function because the integral of $\pi(w)$ over all w is 1. For our training data we also have a loss function, and we seek a value of w so that the loss is minimized. We could express this using the function

$$p(D|w) = \exp(-L(w)^2),$$

where D represents our training data set and $L(w)$ is the loss on training set D for a given value of w. The function $p(D|w)$ is called a likelihood function and the notation indicates that it is the likelihood that we would have the training data D if w were a particular value. In a sense this seems backward: we have our training data and we want to find w. However, we can think about it as what is the likelihood that our model would produce the data in D if w were the value of the parameter. The form of the likelihood above makes it so that the function is maximized when the loss function is small; we could have chosen a function other than the exponential that has this property. The likelihood then says that when the loss is small, the likelihood that our model would produce D is high.

Bayes' rule (or Bayes' theorem) says that given a likelihood function and a prior distribution, we should update the prior distribution to get a *posterior distribution* as the following:

$$\pi(w|D) = \frac{p(D|w)\pi(w)}{\int p(D|w)\pi(w)\,dw}. \tag{1.13}$$

This equation tells us, given the data we have collected, what our new probability distribution for w is. It does not give us a single value for w; rather, it provides a distribution of possible w values. The denominator in Eq. (1.13) is there to make the posterior, $\pi(w|D)$, a properly normalized probability. The integral in the denominator is over the range of possible w values.

In the sense that Bayes' rule gives us a distribution for the parameter w, it is giving us a distribution of machine learning models. This gives us the ability to produce a distribution of predictions in a supervised learning problem. This distribution can be characterized as an uncertainty in the prediction and used to understand where in input space a model is more confident in its prediction.

In more general form we can write Bayes' rule for a vector of parameters, $\mathbf{w}$, as

$$\pi(\mathbf{w}|D) = \frac{p(D|\mathbf{w})\pi(\mathbf{w})}{\int p(D|\mathbf{w})\pi(\mathbf{w})\,\mathbf{w}}. \tag{1.14}$$

Using Bayes' rule to find parameters is different than the approach discussed previously of minimizing the loss function. Finding the values of the parameters that minimize the loss function is equivalent to finding the maximum of the likelihood function used in Bayes' rule. For this reason sometimes minimizing the loss function is called a maximum likelihood method.

It turns out that as difficult as maximizing the likelihood (or minimizing the loss function) is, determining a distribution of parameters using Bayes rule is harder. This is because of the integral in the denominator. In most cases this integral cannot be analytically computed. If the parameter vector is large (as it usually is) performing numerical integration would require an enormous number of evaluations of the loss function at different parameter values, making Bayes' rule impractical for determining distributions of parameters. Thankfully, there have been algorithms developed to sample $\mathbf{w}$ from the posterior distribution without evaluating the denominator. These algorithms are examples of Markov Chain Monte Carlo methods with the Metropolis–Hastings algorithm being the most famous version [4]. These algorithms will provide a set of sampled $\mathbf{w}$ from the posterior distribution that specify a set of machine learning models that can then be used to make a range of predictions. Although knowledge of the uncertainty in a machine learning model is useful, the cost of these sampling methods is typically much greater than maximizing the likelihood.

Bayesian Probability

Bayesian probability gives us a way to express our current expectation for a random variable in terms of a prior distribution and then to use data to update the expectation. Bayesian methods give us a distribution of machine learning models, at the cost of making models harder to train.

1.5   Cross-Validation

Often we are most interested in how a machine learning model will perform on new data to operate that has not been used to train the model. We would like to have a sense of how the model will perform before we use it to make decisions, to operate on its own, or in other critical situations. To approximately quantify the expected performance of the model *faute de mieux* we turn to cross-validation.

In cross-validation for supervised learning we split the available data into a training and test set. The training set is used to set the model parameters (i.e., train the model), and the test set is not used to train the model, rather the model attempts to predict the dependent variable for the cases in the test set. This allows us to see how the model will perform on data it has not seen. We can inspect for overfitting, biases in the model, and other problems. Using the test set we can see what the expected accuracy of the model will be on new data and determine if there is a fundamental difference in the model performance on the training and test data. We can also use cross-validation for unsupervised learning problems to test the results (i.e., association rules, low-dimensional representations, etc.) on data not used to formulate the model and test whether the results of the analysis make sense or are otherwise not extendable.

With splitting the data into a test and training set we are potentially limiting the performance of the machine learning model because more data almost always makes for a better model. To address this, k-fold cross-validation is used. In this case the data is split via random assignment into k equally sized groups of cases; $k - 1$ of these groups are used as training data, and the group left out is the test data. This procedure is repeated k times (implying the k models must be trained) giving k different sets of results on test data. This mitigates the impact of not including all the data to train the model and gives us a larger set of test runs to base our judgment of the models performance on.

The extreme case of k-fold cross-validation is leave-one-out cross-validation where k is equal to the number of cases, that is, $k = I$. In this case, we build all models with all of the cases except one and then try to predict that single case, and this is repeated I times. Leave-one-out cross-validation is useful when the number of cases is small and we cannot afford to lose any data, but still want to know how the model would perform on the next case it would encounter. For large data sets leave-one-out cross-validation may be too costly to implement because it would involve building thousands or millions of models in many instances.

The random assignment to the k groups is important if there was any structure in the data collection that would not be expected to be replicated when the model is deployed. For example, if the data were collected using a technique where only a single independent variable is changed at a time, before then changing several variables, one would not want all of the cases in a group to only have a single independent variable changing. However, random assignment is not always the appropriate method for assigning cases. If we are interested in predicting a time series, the test and training sets must respect the arrow of time: we cannot use data from $t' > t$ in a model to predict the outcome at t. Therefore, the cross-validation procedure should break data into potential groups based on the time it was collected to see how the model would perform in predicting the dependent variables at the next time.

Cross-validation is not just for making inferences about model performance on new data; we can use it to set parameters in the model in some cases. In models that have a penalty term, such as λ in Eqs. (1.5) and (1.6), we can use a training and test data set to get an idea of which values of the penalty appropriately balance accuracy and the penalty. Nevertheless, these models must then undergo further cross-validation to observe their performance on data not used to train the model or set the penalty.

Finally, we mention that cross-validation is not the only step required to make a machine learning model acceptable for deployment. It is sound practice to have a warm start or phased deployment of any model. For instance, one could deploy a machine learning model to give predictions alongside a human operator for a period of time. During that time the human operator may not even know what the machine is predicting, but we can go back and check the accuracy. The next step may be to have the machine make predictions with the human approving or agreeing to the decisions for a period of time. As automatic use of the model grows, the amount of human involvement can then be decreased.

Another important aspect of cross-validation and model deployment is knowing when the machine learning model is uncertain. There are types of models that can automatically indicate how uncertain a given prediction is. This information can be used to highlight decisions for human intervention. Additionally, it can be possible to determine if a case is very different than the cases used to train the model. For example, if the case has a particular value of an independent variable that is an order of magnitude larger than any in the training set, we would be wary of the confidence to place in the model's prediction. It is these cases where human intervention (or post-prediction review) may be necessary.

Cross-Validation

Cross-validation splits the available data into training and test sets in an attempt to determine how the model will perform on new data. The machine learning model is trained using the training data, and then the model's performance is assessed on the test data. This process can be repeated by splitting the data into different sets repeatedly. A common form of cross-validation is k-fold cross-validation where:

- The available data is split randomly into k groups.
- $k - 1$ of the groups are used to build the model, and the remaining group is the test data.
- This process is then repeated k times.
- When the number of groups is equal to the number of cases, we call this leave-one-out cross-validation.

Notes and Further Reading

The topic of combining uncertainty with prediction from machine learning is covered in some detail in a previous monograph by the author [5].

Problems

1.1 Use Bayesian statistics to give the posterior probability of the toss of a particular coin landing on "heads," θ. Use a prior distribution of

$$
\pi(\theta) = \begin{cases} 1 & \theta \in [0, 1] \\ 0 & \text{otherwise} \end{cases},
$$

and for the data likelihood use the probability from a binomial random variable

$$
p(h, n|\theta) = \binom{n}{h} \theta^h (1 - \theta)^{n-h},
$$

where h is the number of heads in n tosses of the coin. Assuming you have flipped the coin 10 times and obtained 6 heads, plot the posterior distribution for θ. Now if you toss the coin 100 times and get 55 heads, how does the distribution change?

1.2 Consider the minimization problem to minimize the function $f(x, y) = (2x + y - c)^2 + (x - y - d)^2$ subject to the constraint $x^2 + y^2 \leq 1$. Use Lagrange multipliers to show that there is an equivalent loss function of the form

$$L = (2x + y - c)^2 + (x - y - d)^2 + \lambda(x^2 + y^2),$$

and determine the value of λ.

1.3 For the softmax function, $\text{softmax}(\ell, z_0, z_2, \ldots, z_L)$, determine the minimum value z_ℓ must have to have a probability greater than 90% as a function of L. What is the minimum value to have a probability greater than 99%?

1.4 Consider the binary cross-entropy loss function. Suppose you have a binary random variable, y, where the probability of a given case y_i being equal to 1 is 0.8. What is the expected value of the loss function for I cases if your model says that $h = 0.8$ for every case?

References

1. Volodymyr Mnih, Koray Kavukcuoglu, David Silver, Andrei A Rusu, Joel Veness, Marc G Bellemare, Alex Graves, Martin Riedmiller, Andreas K Fidjeland, Georg Ostrovski, et al. Human-level control through deep reinforcement learning. *Nature*, 518(7540):529, 2015.
2. David Silver, Aja Huang, Chris J Maddison, Arthur Guez, Laurent Sifre, George Van Den Driessche, Julian Schrittwieser, Ioannis Antonoglou, Veda Panneershelvam, Marc Lanctot, et al. Mastering the game of go with deep neural networks and tree search. *Nature*, 529(7587):484, 2016.
3. J L Peterson, K D Humbird, J E Field, S T Brandon, S H Langer, R C Nora, B K Spears, and P T Springer. Zonal flow generation in inertial confinement fusion implosions. *Physics of Plasmas*, 24(3):032702, 2017.
4. W.R. Gilks and DJ Spiegelhalter. *Markov chain Monte Carlo in practice*. Chapman and Hall/CRC, 1996.
5. Ryan G McClarren. *Uncertainty Quantification and Predictive Computational Science*. Springer, 2018.

Chapter 2
Linear Models for Regression and Classification

And beneath the jazz a cortex, a stiffness or stillness;
The older shell, varnished to lemon colour,
Brown-yellow wood, and the no colour plaster,
Dry professorial talk...

—Ezra Pound, The Seventh Canto

Abstract This chapter discusses linear regression and classification, the foundations for many more complex machine learning models. We begin with a motivating example considering an object in free fall to then use regression to find the acceleration due to gravity. This example then leads to a discussion of least squares regression and various generalizations using logarithmic transforms. The topic of logistic regression is presented as a classification method, with an example given to predict failure of a part based on the temperature it was exposed to. Then the topic of regularization based on the ridge, lasso, and elastic net regression methods is presented as a means to select variables and prevent overfitting. These methods are then used in the case study to determine the laws of motion of a pendulum using data.

Keywords Linear regression · Logistic regression · Regularized regression · Ridge regression · Lasso regression · Interpreting models

2.1 Motivating Example: An Object in Free Fall

Consider the experiment where an object at rest is dropped from a known height. We take an object up to a given height above the ground, call this height h, and drop it. We then measure the time t it takes to reach the ground. Newton's laws will tell us that in the absence of drag, h and t are related by the equation

$$h = v_0 t + \frac{g}{2} t^2, \tag{2.1}$$

© Springer Nature Switzerland AG 2021
R. G. McClarren, *Machine Learning for Engineers*,
https://doi.org/10.1007/978-3-030-70388-2_2

where v_0 is the initial vertical velocity of the object and g is the acceleration due to gravity. Suppose we perform 8 experiments at different values of h and t, and we want to use this data to inform a prediction of the height an object was dropped from given the time it took to reach the ground. That is we wish to find $h(t)$ for our particular experimental setup. We know that in our experiment there will be measurement uncertainty so we do not expect (2.1) to hold exactly; moreover our experiments may have a bias in that the apparatus might systematically drop objects from slightly higher or lower by an amount h_0 that is subtracted from the left-hand side of (2.1). Nevertheless, we would like to use our data to estimate the terms in the model and find any systematic biases in the measurements.

Using (2.1) we have a model form for the function with two unknown constants v_0 and g, and an unknown bias term. If we do not assume that $v_0 = 0$, because we are not sure our experimental apparatus can assure zero initial velocity, we can think of this as a regression problem with input, t, and a single output h. However, the model we are trying to fit has a t^2 term in it, and it offers no complication to call this another input given that, if one knows t, t^2 can be directly computed. We can cast this model in generic form by defining $x_1 = t$, $x_2 = t^2$, $b = -h_0$, and $y = h$ to write

$$y = w_1 x_1 + w_2 x_2 + b. \tag{2.2}$$

This model is shown graphically in Fig. 2.1. The inputs are denoted by the circles on the left. These inputs are connected to the output via the weights w_1 and w_2; the intercept (or bias) also connects to the output. The output node in the figure has Σ inside it to indicate that it computes the sum of the inputs times the weight on the connecting arrow, i.e., this node computes $w_1 x_1 + w_2 x_2 + b$.

Now that our model is specified we can use the data we collected to determine the three free parameters in the model w_1, w_2, and b. To do this we specify a *loss function* for the model. Given that we have performed 8 measurements and recorded y_i and $(x_1)_i$, $(x_2)_i$ for $i = 1, \ldots, 8$, we would like our model to minimize the difference between $w_1 (x_1)_i + w_2 (x_2)_i + b$ and y_i. One way to do this would be to minimize the sum

Fig. 2.1 Schematic of the model $y = w_1 x_1 + w_2 x_2 + b$. The inputs to the model are x_1 and x_2, the intercept (or bias) is b, and the output is y. The labels on the connecting lines indicate the multiplicative weight of the connection. The Σ in the output node indicates that this node computes the sum of its inputs

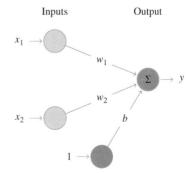

$$L = \sum_{i=1}^{8} (\hat{y}_i - y_i)^2, \tag{2.3}$$

where $\hat{y}_i \equiv w_1(x_1)_i + w_2(x_2)_i + b$ is the predicted value for the model for case i. We choose the sum of the squares for two reasons: each term in the sum is positive so there will not be cancellation of errors (if one term is positive and another negative, we could get a small sum even if each error was large), and it is a differentiable function. The function L is called the *loss function* for our model. Other loss functions could be used (e.g., the sum of the absolute values of error), but the sum of the squared errors is easy to work with.

To find the minimum of L we will need to take the derivative with respect to each of the parameters w_1, w_2, and b and set the result of each to zero. This gives us the three equations

$$\frac{\partial L}{\partial w_1} = \sum_{i=1}^{8} 2(\hat{y}_i - y_i)(x_1)_i = 0, \tag{2.4a}$$

$$\frac{\partial L}{\partial w_2} = \sum_{i=1}^{8} 2(\hat{y}_i - y_i)(x_2)_i = 0, \tag{2.4b}$$

$$\frac{\partial L}{\partial b} = \sum_{i=1}^{8} 2(\hat{y}_i - y_i) = 0. \tag{2.4c}$$

We can rearrange these equations by defining the matrix of dimension 8×3

$$\mathbf{X} = \begin{pmatrix} 1 & x_{11} & x_{12} \\ 1 & x_{21} & x_{22} \\ \vdots & \vdots & \vdots \\ 1 & x_{81} & x_{82} \end{pmatrix},$$

and vectors $\mathbf{w} = (b, w_1, w_2)^{\mathrm{T}}$, $\mathbf{y} = (y_1, y_2, \ldots, y_8)^{\mathrm{T}}$, and x_{ij} being the value of $(x_j)_i$. With these definitions $\hat{\mathbf{y}} = (\hat{y}_1, \ldots, \hat{y}_8)^{\mathrm{T}} = \mathbf{X}\mathbf{w}$, and we can write the system in (2.4) as

$$\mathbf{X}^{\mathrm{T}}\mathbf{X}\mathbf{w} = \mathbf{X}^{\mathrm{T}}\mathbf{y}.$$

Notice that $\mathbf{X}^{\mathrm{T}}\mathbf{X}$ will give a square matrix so in principle this equation can have a unique solution. In practice as long as the number of measurements is larger than the data and the rows are not all linearly dependent, there will be a solution.

To finish the example we specify the data from the experiments. For the 8 measurements we have

$$\mathbf{h}[m] = \begin{pmatrix} 10.1 \\ 25.7 \\ 17.5 \\ 12.3 \\ 9.8 \\ 26.7 \\ 28.4 \\ 21.0 \end{pmatrix}, \qquad \mathbf{t}[s] = \begin{pmatrix} 1.46 \\ 2.3 \\ 1.9 \\ 1.6 \\ 1.43 \\ 2.35 \\ 2.42 \\ 2.08 \end{pmatrix}.$$

These correspond to

$$\mathbf{X} = \begin{pmatrix} 1 & 1.46 & 2.12 \\ 1 & 2.3 & 5.3 \\ 1 & 1.9 & 3.62 \\ 1 & 1.6 & 2.56 \\ 1 & 1.43 & 2.06 \\ 1 & 2.35 & 5.5 \\ 1 & 2.42 & 5.85 \\ 1 & 2.08 & 4.34 \end{pmatrix}, \qquad \mathbf{X}^T\mathbf{X} = \begin{pmatrix} 8 & 15.54 & 31.35 \\ 15.54 & 31.35 & 65.3 \\ 31.35 & 65.3 & 139.74 \end{pmatrix}, \qquad \mathbf{X}^T\mathbf{y} = \begin{pmatrix} 151.5 \\ 315.94 \\ 676.64 \end{pmatrix}.$$

From this data we find that $w_1 = 4.905$, $w_2 = 0.00954155$, and $b = -0.30071811$. Using the definitions above this gives that $g \approx 9.81$ m/s^2, $v_0 \approx 0.01$ m/s, and $h_0 \approx 0.3$ m.

2.2 General Linear Model

The example above can be extended to J independent variables x_j for $j = 1, \ldots, J$ and single dependent variable y as

$$y = \sum_{j=1}^{J} w_j x_j + b, \tag{2.5}$$

where, as before, b is the bias in the model. In this model the matrix $\mathbf{X}$ is of size $I \times (J + 1)$, where I is the number of observations and

$$\mathbf{X} = \begin{pmatrix} 1 & x_{11} & x_{12} & \ldots & x_{1J} \\ 1 & x_{21} & x_{22} & \ldots & x_{2J} \\ \vdots & \vdots & & & \vdots \\ 1 & x_{I1} & x_{I2} & \ldots & x_{IJ} \end{pmatrix}. \tag{2.6}$$

Additionally, the vector $\mathbf{y} = (y_1, \ldots, y_I)^T$ contains the I observations of the dependent variable. As before we find the weights and bias via solving the system

$$\mathbf{X}^T \mathbf{X} \mathbf{w} = \mathbf{X}^T \mathbf{y}, \tag{2.7}$$

with $\mathbf{w} = (b, w_1, \ldots, w_J)^T$.

The resulting values of b and w_j are the values that minimize the sum of the squared-error loss function given by

$$L = \sum_{i=1}^{I} (\hat{y}_i - y_i)^2, \tag{2.8}$$

where $\hat{y}_i$ is the prediction of the model for case i. For this reason the resulting linear model is known as the least squares linear regression model or just least squares (LS) model. This is the simplest linear model that can be defined. There has been much research into when such models work and how to diagnose them. It has been shown that these models work best in practice when the error in the model is independent of the value of the prediction, i.e., the difference $\hat{y}_i - y_i$ does not depend on the magnitude of y_i. This can be checked by creating a scatter plot of $\hat{y}$ versus $\hat{y} - y$ and looking for an obvious relationship between the two.

We can visualize the linear model as a mapping from inputs to an output in Fig. 2.2. This figure represents the inputs as nodes at the left that are connected with weights w_j to a node that sums its inputs to produce the output y. The bias is connected to the summation node with a weight of b.

One feature of linear models is that they lend themselves to interpretation. The w_j tells us how much we expect y to change for a unit change of x_j, if every other variable stays the same. Thinking about it another way, w_j is the derivative of y with respect to x_j. This can be very useful to explain what is driving the dependent

Fig. 2.2 Schematic of the general linear model $y = w_1 x_1 + w_2 x_2 + \cdots + w_J x_J + b$. There are J inputs to the model. The labels on the connecting lines indicate the multiplicative weight of the connection. The Σ in the output node indicates that this node computes the sum of its inputs

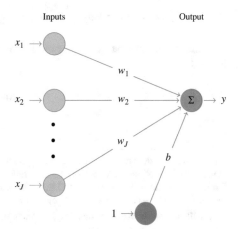

variable. Care must be taken, however, if the model includes independent variables that are transformations of each other. In the example we had a $x_1 = t$ and $x_2 = t^2$: in such a scenario it is not possible to keep x_2 fixed while adjusting x_1. This scenario would require transforming back to the original t variable and then taking the derivative.

We can also interpret the bias, b, as the value y would take if all of the independent variables were zero. Sometimes this can cause consternation among non-technical stakeholders because it may not make sense (or even be possible) for all of the independent variables to be zero in the model.[1] It is possible to remove the bias from the model (by removing the column of ones in the data matrix $\mathbf{X}$). Alternatively, one can center the inputs by defining new variables $\tilde{x}_j \equiv x_j - \bar{x}_j$, where $\bar{x}_j$ is the mean of the jth independent variable. In this case, the bias would be the predicted response when all the inputs were at their average, or mean, value.

2.2.1 Nonlinear Transformations

We have already seen that nonlinear transformations of the independent variables are possible in the model when we included t^2 in our example model in Sect. 2.1. Other transformations are possible. For instance, we could define a new independent variable as the cosine of another independent variable; such a transformation is important if there is expected periodic response to an input. As mentioned above, care must be taken when interpreting models with these transformations, but fitting them is straightforward.

Exponential Models
The dependent variable can also be transformed to express different behaviors. Logarithmic transformations are useful in this regard, especially if the dependent variable is a strictly positive number. Consider an *exponential* model of the form

$$y = \exp[w_1 x_1 + \cdots + w_J x_J + b] = e^{\sum_{j=1}^{J} w_j x_j + b}. \qquad (2.9)$$

This model can be cast as a linear model by defining a new variable using the natural logarithm[2]

$$z = \log y. \qquad (2.10)$$

---

[1] A personal anecdote regarding this. I was explaining to an executive why a linear model predicted that the number of people who would visit a retail store would be negative if no one lived within 5 miles of the store. There was no training data with fewer than 10,000 people living within that radius; the model has no idea what is going to happen in that scenario.

[2] We will use the convention that $\log x$ denotes the natural logarithm of x. When we need the logarithm in base 10, we use $\log_{10} x$.

Therefore, we can take the natural logarithm of both sides of Eq. (2.9) to get

$$z = \sum_{j=1}^{J} w_j x_j + b, \tag{2.11}$$

which is a general linear model of the form Eq. (2.5) with a dependent variable given by $z = \log y$. Therefore, to fit this model we only need to interpret the dependent variable as the logarithm of the original dependent variable. When fitting the model, we will be minimizing the sum of $(\hat{z}_i - z_i)^2$, that is, minimizing the sum of the squared difference in the logarithm of the dependent variable. Using the properties of the logarithm, we find

$$\hat{z}_i - z_i = \log\left(\frac{\hat{y}_i}{y_i}\right) \qquad i = 1, \ldots, I.$$

Therefore, we can think of the model as minimizing the error on a multiplicative scale: the model wants the ratio of $\hat{y}_i$ to y_i to be as close to 1 as possible.

Interpretation of exponential models is a bit different than purely linear models. In an exponential model, when x_j is increased the effect is multiplicative. Consider a case where all of the x_j are independent. The derivative of z with respect to x_j is

$$\frac{\partial z}{\partial x_j} = w_j.$$

We also use the definition of z to get the derivative of y with respect to x_j:

$$\frac{\partial z}{\partial x_j} = \frac{\partial}{\partial x_j} \log y = \frac{1}{y} \frac{\partial y}{\partial x_j}.$$

Therefore, for the exponential model

$$\frac{\partial y}{\partial x_j} = w_j y. \tag{2.12}$$

This implies that as the value of the dependent variable increases, the multiplicative effect of increasing x_j also increases.

A useful rule of thumb can be found by supposing a small change to x_j and using Taylor series to find the change in y

$$\frac{y(x_j + \delta) - y(x_j)}{y(x_j)} = \delta w_j + \frac{1}{2}\delta^2 w_j^2 + \frac{1}{6}\delta^3 w_j^3 + \cdots \approx \delta w_j. \tag{2.13}$$

Therefore, if we increase x_j by δ, we change the prediction by a factor of $1 + w_j\delta$. Suppose that w_j is 0.1, this would imply that increasing x_j by 1 would approximately increase the value of y by 0.1 or 10%.

To interpret the bias term, we note that when all the independent variables are zero, the prediction for y will be e^b. For this reason, sometimes the exponential model is written as

$$y = ae^{\sum_{j=1}^{J} w_j x_j},$$

where $a = e^b$.

Fitting Exponential Models

To fit a model of the form $y = a\exp(\sum_{j=1}^{J} w_j x_j)$, compute an intermediate variable $z = \log y$ and fit a standard linear model of the form $z = \sum_{j=1}^{J} w_j x_j + b$ using least squares. The weights from the linear model are the coefficients in the exponential model, and $a = e^b$.

Power-Law Models

It is also possible to fit power-law models of the form

$$y = at_1^{w_1} t_2^{w_2} \cdots t_J^{w_J} = a \prod_{j=1}^{J} t_j^{w_j}. \tag{2.14}$$

To formulate the power-law model as a linear model we make the transformations

$$z = \log y, \qquad x_j = \log t_j, \qquad b = \log a.$$

Using these definitions we take the logarithm of both sides of Eq. (2.14)) to get

$$z = b + w_1 \log t_1 + w_2 \log t_2 + \cdots + w_J \log t_J = \sum_{j=1}^{J} w_j x_j + b. \tag{2.15}$$

Therefore, we can transform a power law into a linear model by taking the natural logarithm of both the independent and dependent variables.

Fitting Power-Law Models

To fit a model of the form $y = a \prod_{j=1}^{J} t_j^{w_j}$, compute intermediate variables $z = \log y$ and $x_j = \log t_j$ and fit a standard linear model of the form $z =$

(continued)

$\sum_{j=1}^{J} w_j x_j + b$ using least squares. The weights from the linear model are the coefficients in the power-law model, and $a = e^b$.

2.3 Logistic Regression Models

There is a particular transformation for linear models known as the logistic transform that is used in classification problems. We will first concern ourselves with binary classification, that is, an observation of class 0 or class 1.

As a motivating example, we consider a hypothetical data set that contains information for support parts from a structure. The parts were each exposed to a heat load of different magnitudes, and we have recorded whether the part failed or not: parts that failed are in class 1 and those that did not fail are in class 0.

We denote the load as x and the class as $y \in \{0, 1\}$. We could use the linear models discussed in the previous section to formulate a model of the form $\Pr(y = 1|x) = wx + b$, where $\Pr(y = 1|x)$ is the probability that $y = 1$ given a value of x. Such a model would suffer from many drawbacks, perhaps the most obvious being the fact that in reality the probability must be between 0 and 1 and there is no way to enforce such a condition in the model.

Rather than predicting the probability directly, we will make a model to give the logarithm of the odds of $y = 1$. These *log-odds* are defined as

$$z = \log \left(\frac{\Pr(y = 1|x)}{1 - \Pr(y = 1|x)} \right). \tag{2.16}$$

The log-odds map the probability from the interval $[0, 1]$ to the real line, $(-\infty, \infty)$. The log-odds are also invertible to find the probability. If z is the log-odds of $y = 1$ given x, then

$$\Pr(y = 1|x) = \frac{1}{1 + e^{-z}} \equiv \text{logit}(z). \tag{2.17}$$

The function $\text{logit}(z)$ is known as the logistic function, and it is this function that gives logistic regression its name. The probability of $y = 1$ as a function of the log-odds is shown in Fig. 2.3. From the figure we can see that if the log-odds are zero (or the odds are 1), then there is an equal probability that the observation is in class 1 or 0. As z approaches infinity, the probability approaches 1; similarly, the probability approaches zero as z goes to negative infinity.

We write a linear model for the log-odds of the form

Fig. 2.3 Plot of the
probability that an
observation is in class $y = 1$
as a function of the log-odds.
The value of $\Pr(y = 1)$ at
$z = 0$ is 0.5. At $z = 2$,
$\Pr(y = 1) \approx 0.88$; at $z = -2$,
$\Pr(y = 1) \approx 0.12$

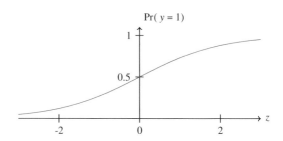

$$\log\left(\frac{\Pr(y = 1|x)}{1 - \Pr(y = 1|x)}\right) = wx + b. \tag{2.18}$$

To fit this model we need to find the weight, w, and bias, b, that minimize a loss function. The loss function we use should go to zero as the number of correct class predictions goes up. To this end we assume that at a given value of the independent variable x there is a unique probability of an observation having $y = 1$ using Eq. (2.18). Therefore, for a given observation, the model predicts

$$\Pr(y|x) = h(x)^y(1 - h(x))^{1-y},$$

where $h(x) = 1/(1 + \exp(wx + b))$. By inspection this formula indicates that to maximize the probability of the model being correct when $y = 1$, we want $h(x) \rightarrow 1$; conversely, if $y = 0$, we want $h(x)$ to go to zero to be correct.

Now consider we have I observations y_i with independent variable x_i. We want to maximize the probability that the model is correct on all of the observations. In other words, we want to maximize the product

$$\prod_{i=1}^{I} h(x_i)^{y_i}(1 - h(x_i))^{1-y_i}.$$

This product would be 1 if the model were perfect. We define the loss function as the negative logarithm of this product (the negative is chosen to make the optimal solution at the minimum of the loss function):

$$L = -\sum_{i=1}^{I} \log\left[h(x_i)^{y_i}(1 - h(x_i))^{1-y_i}\right]. \tag{2.19}$$

This loss function does not lead to a neat, closed form solution like the least squares model did. Nevertheless, the minimum of the loss function can be found using standard techniques such as gradient descent.

Returning to our example, suppose we have the following 10 observations ($I = 10$):

T(°C)	281	126	111	244	288	428	124	470	301	323
Result (Fail=1)	1	0	0	0	0	1	0	1	0	1

Fig. 2.4 Comparison of the
example data with the
predicted probability of
failure from the logistic
regression model

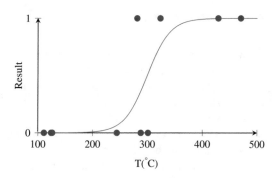

From this data we find the optimal logistic model is

$$z = 0.04831T - 14.46073.$$

This model predicts that there is a 50% probability of failure at $T = 299.3\,°\mathrm{C}$.
When we look at the model and the data in Fig. 2.4, we see that the point where
the predicted probability rapidly transitions from a very low likelihood of failure
to a high likelihood occurs in the temperature region where there are cases of both
failure and no failure.

2.3.1 Comparing to the Null Model

One question we might naturally ask is how much better is our model than random
chance. In our example we have 4 failures and 6 successes. Therefore, we could say
that any case has a 40% chance of being a failure, regardless of the temperature.
Using this chance we could predict that no case is expected to fail and be right 60%
of the time. This is the so-called null model where we say temperature has no effect
on failure. In this case the loss function is

$$L_{\mathrm{null}} = -\sum_{i=1}^{I} \log\left[\theta_i^y (1 - \theta)^{1-y_i}\right],$$

where θ is the fraction of cases that $y = 1$ (0.4 in our example).

We can compare the loss of the null model with the loss of the fitted model.
In the example, using $\theta = 0.4$ and the model fitted above, we get a null loss of
$L_{\mathrm{null}} = 6.73$ and a loss of $L = 2.77$ for the fitted model. This decrease in the loss
function by over half indicates that the addition of the temperature variable to the
prediction gives a significant improvement in the predictive power.

2.3.2   Interpreting the Logistic Model

Another relevant question is how to interpret the weight and bias in the model. The bias tells us what the probability of $y = 1$ is when $x = 0$ via the formula $1/(1+e^{-b})$. In the example, $b \approx -14.46$, making $\Pr(y = 1|T = 0) \approx 5 \times 10^{-7}$ indicating that there is little chance that at zero temperature we will observe failure.

The weight is a bit trickier to interpret. Clearly, every unit increase in the independent variable increases z by w. The effect on the probability is trickier due to the nonlinear relationship between z and the probability. A useful reference is the "divide by four rule" to get an estimate of the maximum impact of the variable on the probability. We compute the derivative of the probability of $y = 1$ with respect to x as

$$\frac{\partial}{\partial x}\left(\frac{1}{1 + e^{-(wx+b)}}\right) = \frac{we^{-(wx+b)}}{\left(1 + e^{-(wx+b)}\right)^2}. \tag{2.20}$$

To maximize this derivative, and therefore maximize the sensitivity of the probability to x, we take the derivative of Eq. (2.20) and set it to zero and find that the maximum sensitivity for $\Pr(y = 1|x)$ occurs where $z = 0$. This can be seen visually in Fig. 2.3. At $z = 0$ Eq. (2.20) becomes

$$\frac{\partial}{\partial x}\left(\frac{1}{1 + e^{-(wx+b)}}\right)\Bigg|_{z=0} = \frac{w}{\left(1 + e^0\right)^2} = \frac{w}{4}. \tag{2.21}$$

Therefore, a unit increase in x results in a $w/4$ increase in the probability near $z = 0$. This gives us an upper bound on the impact a unit change in the independent variable will have on the probability.

Applying the divide by four rule to the example, we get that near $z = 0$ (that is $T = 299.3\,°\mathrm{C}$), an increase of $1°\mathrm{C}$ increases the probability of failure by $0.04831/4 \approx 0.012$ or 1.2%. We could also say that at this point, a $10°\mathrm{C}$ increase would increase the probability of failure by less than 12%.

2.3.3   Multivariate Logistic Models

Thus far we have discussed logistic classification models with a single independent variable. There is no major complication in extending them to multiple variables. Given J independent variables we write the model as

$$z = \sum_{j=1}^{J} w_j x_j + b, \qquad \Pr(y = 1|x) = \frac{1}{1 + e^{-z}}. \tag{2.22}$$

Fig. 2.5 Schematic of the logistic model for classification $z = w_1x_1 + w_2x_2 + \cdots + w_Jx_J + b$, and $\Pr(y = 1|x) = 1/(1 + \exp(-z))$. There are J inputs to the model. The labels on the connecting lines indicate the multiplicative weight of the connection. The "logit" in the output node indicates that this node computes the logistic function on the sum of its inputs

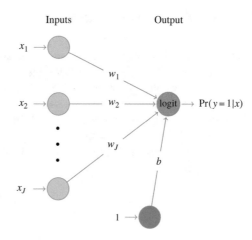

This model is visualized in Fig. 2.5. From the figure we can see that the only difference between the linear regression model and the logistic model is the transformation in the output layer.

In the multivariate case, the loss function is the same as Eq. (2.19). To fit a multivariate model the loss function is minimized with respect to the bias and the J weights. The coefficients can be interpreted using the divide by four rule, as before, and the interpretation of the bias is identical to the single variable case.

Although logistic models are simple models, they can be powerful tools in making predictions. Due to their ease of fitting and evaluation, these models can be deployed on large data sets in a variety of settings. For example, in a manufacturing process logistic regression could be used to understand what conditions lead to products failing quality assurance tests, and we could use this insight to improve operations.

We also note that the logistic model is our first neural network in the sense that it uses a nonlinear thresholding function (in this case the logistic function) to give a nonlinear response to an affine transformation of the inputs.[3] We will return to this idea later.

> **Logistic Regression Models**
> To predict a binary outcome, logistic regression models can be fit to predict the probability that a given case is in class 1 versus class 0 as a function of the independent variables. The basis for logistic regression models is similar to linear regression, but the interpretation of the coefficients is different. The "divide by 4" rule can aid in interpretation.

---

[3] By affine transformation of the inputs, we mean a linear combination of the inputs plus a constant term. The bias supplies the constant term and the weights define the linear combination.

2.3.4 *Multinomial Models*

Logistic models can be extended to multiclass classification problems in what is known as a multinomial classification model. This is for problems where we want to know if an observation will fall into one of several possible classes, i.e., $y \in \{1, 2, \ldots, K\}$. Multinomial regression models are powerful tools that can be quite accurate. Using the right data and model formulation these models can predict the winners at the Academy Awards [1].

To produce this model, we construct models for the log-odds of each outcome without a bias:

$$z^{(1)} = \sum_{j=1}^{J} w_j^{(1)} x_j, \qquad (2.23)$$

$$z^{(2)} = \sum_{j=1}^{J} w_j^{(2)} x_j,$$

$$\vdots$$

$$z^{(K)} = \sum_{j=1}^{J} w_j^{(K)} x_j.$$

Using the fact that the sum of the probabilities must be 1 ($\Pr(y = 1) + \cdots + \Pr(y = K) = 1$), we define the probability that $\Pr(y = c)$ as

$$\Pr(y = c | x) = \frac{e^{z^{(c)}}}{\sum_{k=1}^{K} e^{z^{(k)}}} \equiv \text{softmax}(c, z^{(1)}, \ldots, z^{(K)}), \qquad c = 1, 2, \ldots, K.$$
$$(2.24)$$

Equation (2.24) defines the **softmax** function. This function is named because of the K variables $z^{(c)}$, the one that is the largest will cause the softmax function to be close to one and the others will return a number close to zero. Suppose that $K = 3$ and the $z^{(c)}$ are 0, 2, and 4 for $c = 1, 2, 3$. The values of the softmax function will be

$$\text{softmax}(1, 0, 2, 4) = \frac{1}{1 + e^2 + e^4} \approx 0.016,$$

$$\text{softmax}(2, 0, 2, 4) = \frac{e^2}{1 + e^2 + e^4} \approx 0.117$$

$$\text{softmax}(3, 0, 2, 4) = \frac{e^4}{1 + e^2 + e^4} \approx 0.867.$$

These values sum to one, and the largest number (4 in this case) gave a result closest to one.

The loss function used in fitting must also be changed to deal with the fact that there are multiple categories:

$$L = -\sum_{i=1}^{I}\sum_{k=1}^{K}\mathcal{I}(k, y_i)\log \text{softmax}(k, z^{(1)}, \ldots, z^{(K)}), \qquad (2.25)$$

where

$$\mathcal{I}(k, y_i) = \delta_{k,y_i} = \begin{cases} 1 & y_i = k \\ 0 & \text{otherwise} \end{cases}$$

is known as the indicator function. The loss function defined in Eq. (2.25) is called the cross-entropy function. This function is a measure of the distance between the model's predictive distribution and the distribution of the data (i.e., the empirical distribution).

The multinomial model is shown in Fig. 2.6. The fact that there are more weights is apparent in the figure, but the overall architecture of the model is similar to the logistic regression model with the logit node replaced by a softmax node.

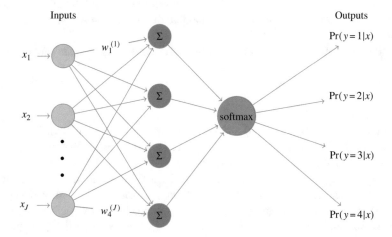

Fig. 2.6 Schematic of the multinomial classification model in the case of $K = 4$ classes. There are J inputs to the model, independent variables. Only two of the $4J$ weights are labeled but each line between the input layer and the "Σ" layer has a weight. As before, the Σ denotes that the sum of the inputs to the node is computed. The softmax function computes the probability of each class that is then returned in the output layer

Multinomial Regression Models

For multiclass problems, where we want to know which of many possible classes a case belongs to, we fit several models that have a similar form to the logistic regression model. We use the softmax function to assure that the probabilities given by each of the logistic regression models sum to 1. Also, we use the more general cross-entropy function as the loss. Nevertheless, working with multiclass problems is not much more difficult than dealing with single-class classification problems.

2.4   Regularized Regression

In the previous sections we detailed several different models for making predictions that were variations on the theme of linear, least squares regression. All of the models had a loss function related to the model error, and fitting the models required minimizing that error. However, minimizing the error is not our only objective as we demonstrate now. Consider a variable y that is uniformly, randomly distributed between 0 and 1. We have two samples of y given by $y_1 = 0.235$ and $y_2 = 0.642$. We then produce samples from another random number x that are $x_1 = 0.999$, and $x_2 = 0.257$.

Using this data we construct a linear regression model using least squares of the form

$$y = wx + b.$$

From the data we find $w = -0.5485$ and the bias to be $b = 0.7830$. This model is exact in that $y_i = wx_i + b$ for $i = 1, 2$. However, recall that y and x were independent random numbers so there should be no relation between them. We see the drawback of this model when we consider a third set of randomly-sampled values of y and x: $y_3 = 0.610$, $x_3 = 0.962$. In this case the model predicts $\hat{y}_3 = 0.255343$, which is off by over 300%—so much for the perfect model. As we continue sampling y and x points, the model's predictive ability goes to zero. This phenomenon is known as overfitting.

Overfitting can also occur when we have many data points. We can "improve" the model by adding variables to the model to make the model's error on the known data go to zero. For instance, if we have N realizations of the dependent variables, then a model with N independent variables, even if some of those independent variables are meaningless noise, will be perfect on the known data. Nevertheless, on new data the model will have very little predictive power (see Sect. 1.1.1.1 for a demonstration). This is part of the motivation for cross-validation discussed in the previous chapter.

There are also occasions where we have the opposite problem: we have more independent variables than realizations. If we consider an experiment where the number of controllable parameters in the experiment is J but the number of realizations of the experiment we can perform is I with $I < J$, how can we decide which of the independent variables to include in a model when we assume that only a few of the independent variables affect the dependent variables?

The idea of "regularized regression" attempts to address both the problem of overfitting to the known data *and* variable selection. The idea is to change the loss function to balance the objectives of minimizing the error with reducing the magnitude of the weights and biases in the model. If the model prefers smaller in magnitude weights and biases, the model can be robust to overfitting and help to select variables as we will see.

2.4.1 Ridge Regression

Ridge regression [2], also known as Tikhonov regularization, adjusts the loss function by adding the Euclidean norm, also called the L_2 norm, of the weights and biases. For linear regression the loss function becomes

$$L_{\text{ridge}} = \sum_{i=1}^{I} (\hat{y}_i - y_i)^2 + \lambda \|\mathbf{w}\|_2, \tag{2.26}$$

where, as before, $\hat{y} = w_1 x_1 + \dots w_J x_J + b$, I is the number of observations, and J is the number of independent variables. We have written the weights and biases in the vector $\mathbf{w} = (b, w_1, \dots, w_J)^{\mathrm{T}}$ and defined the L_p norm for a vector of size M to be

$$\|\mathbf{x}\|_p = \left(\sum_{\ell=1}^{M} |x_\ell|^p \right)^{1/p}. \tag{2.27}$$

Notice that $p = 2$ is the Euclidean norm: $\sqrt{x_1^2 + \dots x_M^2}$.

In the ridge loss function the parameter $\lambda \geq 0$ is the penalty strength: it gives the relative weight of the penalty to the error when minimizing the loss function. When $\lambda = 0$, the loss function becomes the same as the least squares model.

For $\lambda > 0$ ridge regression will give a result even in the case where $J > I$. As a consequence ridge regression will allow us to fit a line given a single point! To see that ridge regression will give a unique solution when $J > I$ we take the derivative of Eq. (2.26) with respect to the J weights and the bias and set the result to 0 to get a system of $J + 1$ equations for the weights and bias:

$$(\mathbf{X}^{\mathrm{T}}\mathbf{X} + \lambda \mathbf{I})\mathbf{w} = \mathbf{X}^{\mathrm{T}}\mathbf{y}, \tag{2.28}$$

where $\mathbf{I}$ is an identity matrix of size $(J+1) \times (J+1)$, and the other matrices and vectors are the same as in Eq. (2.7). Notice that this equation will have a solution when $\lambda > 0$ because the penalty adds a constant to the diagonal (i.e., the ridge) of the matrix $\mathbf{X}^\mathsf{T}\mathbf{X}$. Therefore, if the rank of $\mathbf{X}^\mathsf{T}\mathbf{X}$ is less than $J+1$, the penalty fixes the rank.

To further understand what the ridge penalty does, we change the problem from minimizing the loss function to a constrained optimization problem. If we think of λ as a Lagrange multiplier we can form the equivalent minimization problem to minimizing L_{ridge} as

$$\mathbf{w}_{\text{ridge}} = \min_{\mathbf{w}} \sum_{i=1}^{I} (\hat{y}_i - y_i)^2 \quad \text{subject to} \quad \|\mathbf{w}\|_2 \le s.$$

There is a one-to-one relationship between s and λ, but we will not derive this here (this was given as an exercise in the previous chapter). We can say that as $\lambda \to 0$, s will go to infinity, implying there is no constraint on the weights.

To illustrate the effect of the ridge penalty, consider a system with $J = 2$ as shown in Fig. 2.7. The loss function in Eq. (2.26) is a quadratic function in the two parameters b and w_1. The contours of this quadratic function will appear as ellipses in the (b, w_1) plane. In the center of these ellipses is the least squares estimate.

The circle in Fig. 2.7 has a radius s, and the solution must lie inside or on this circle. Because the LS estimate is outside the circle, the solution will be where the circle intersects a contour line of the sum of the squared errors at the minimal

Fig. 2.7 Depiction of the ridge regression result $y = w_1 x + b$ compared with least squares. The ellipses are the surfaces of equal value of the sum of the squared errors in the regression estimate. Given that the error has a quadratic form, the ellipses further away from $\hat{\mathbf{w}}_{\text{LS}}$ have a larger error. The circle is $b^2 + w_1^2 = s^2$. The ridge regression solution occurs where an ellipse touches the circle

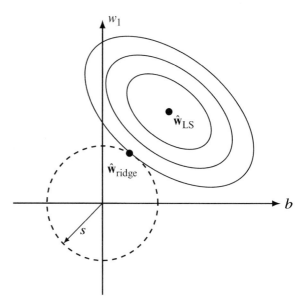

possible value of the error. Notice that the magnitude of both b and w_1 has decreased in the ridge estimate compared with the LS estimate, and that both are non-zero.

A feature of ridge regression is that the larger the value of λ, the smaller the values in $\hat{\mathbf{w}}_{\text{ridge}}$ will be in magnitude relative to the values of $\hat{\mathbf{w}}_{\text{LS}}$, when the LS values exist. This makes λ a free parameter that must be chosen based on another consideration. We will discuss using cross-validation to choose this parameter later.

As a simple example of ridge regression, consider the problem of estimating a function of the form $y = w_1 x + b$ given the data $y(2) = 1$. That is, we are interested in fitting a line to single data point. This problem is formulated as

$$\mathbf{X} = (1, 2), \qquad \mathbf{w} = (b, w_1)^{\text{T}}, \qquad \mathbf{y} = 1.$$

Using these values in Eq. (2.28) we get

$$\begin{pmatrix} 1+\lambda & 2 \\ 2 & 4+\lambda \end{pmatrix} \begin{pmatrix} b \\ w_1 \end{pmatrix} = \begin{pmatrix} 1 \\ 2 \end{pmatrix}.$$

The solution to this equation is

$$w_1 = \frac{2}{\lambda + 5}, \qquad b = \frac{1}{\lambda + 5},$$

for $\lambda > 0$. From this we can see that the limit of this solution as λ approaches zero from the above is

$$\lim_{\lambda \to 0^+} w_1 = \frac{2}{5}, \qquad \lim_{\lambda \to 0^+} b = \frac{1}{5}.$$

Notice that for $\lambda > 0$, the fitted solution does not pass through the data, that is, $2w_1 + b \neq 1$. Also, we note that it is possible to take the ridge solution in the limit of $\lambda \to 0$, but setting $\lambda = 0$ in Eq. (2.28) does not yield a solution.

In this example we can see that we can fit a line to single data point, but the result is not necessarily faithful to the original data, though it does give us a means to fit a solution when $I < J$. This property will be useful for estimating local sensitivities when we have fewer cases than parameters.

2.4.2   Lasso and Elastic Net Regression

If we change the loss function that we used in the ridge regularization to be the L_1 norm of the weights and biases, that is if we include the sum of the absolute values of the w_j's and b, we get the method known as least absolute shrinkage and selection operator, or the "lasso" for short [3]. The loss function for lasso is

$$L_{\text{lasso}} = \sum_{i=1}^{I} (\hat{y}_i - y_i)^2 + \lambda \|\mathbf{w}\|_1 \tag{2.29}$$

$$= \sum_{i=1}^{I} (\hat{y}_i - y_i)^2 + \lambda |b| + \sum_{j=1}^{J} \lambda |w_j|.$$

This seemingly small change has an important impact on the resulting method. For instance, the equations are now more difficult to solve because the absolute value function is not differentiable when its argument is zero. Additionally, the resulting **w** found by lasso tends to set a subset of the elements of this vector to zero. It "lassos" the important independent variables. In this sense, it is a "sparse" method in that lasso can find a solution where only a subset of the independent variables contributes to the model prediction. One feature of the lasso result is that for I data points used to train the model, at most I coefficients of **w** will be non-zero.

We can see how lasso sets some coefficients to zero by investigating the solution for $J = 2$. In this case, level curves in the (b, w_1) plane with a constant value of $\|\mathbf{w}\|$ will be diamond-shaped rather than circular; this is shown in Fig. 2.8. Given this diamond shape the ellipses that represent the curves of equal squared error in the loss function will likely touch the diamond at one of its vertices. These vertices correspond to only one of the coefficients having a non-zero value. In the figure we also notice that the prediction error from lasso will be larger than that for ridge

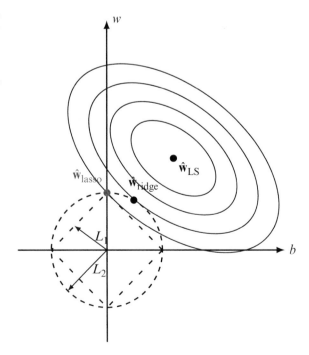

Fig. 2.8 Comparison of lasso and ridge regression result for $y = w_1 x + b$ compared with least squares. The diamond shape is the curve $|b| + |w_1| = s$. With lasso the solution is where the ellipse touches the diamond

because the ellipse that touches the diamond is further away from the least squares result.

We can also combine lasso and ridge loss functions to get the elastic net method [4]. In an extension of the metaphor of the method capturing the important coefficients, this time we use a net rather than a rope. The elastic net loss function is

$$L_{el} = \sum_{i=1}^{I} (\hat{y}_i - y_i)^2 + \lambda \left(\alpha \|\mathbf{w}\|_1 + (1 - \alpha) \|\mathbf{w}\|_2^2 \right), \tag{2.30}$$

with $\alpha \in [0, 1]$. This loss function can transition from lasso to ridge regression: by setting $\alpha = 1$ to get lasso, $\alpha = 0$ to get ridge, and a mixture of the two for α in between 0 and 1. The elastic net allows more than I coefficients to be non-zero and demonstrates the grouping effect: independent variables that are correlated have weights that are nearly identical. The relaxation of the strictness of the lasso allows elastic net to outperform lasso in many problems in terms of minimizing the error in the predicted values of the dependent variables.

In our visualization of the case of $J = 2$, the effect of elastic net is to blunt points of the diamond-shaped level curve of the lasso method. As shown in Fig. 2.9 the elastic net solution is between the ridge and lasso solutions.

One key feature of regularized regression techniques is that the scale of the variable matters. In standard regression if we multiply an independent variable by a constant, it does not change the predictions of the model: the coefficients will adjust

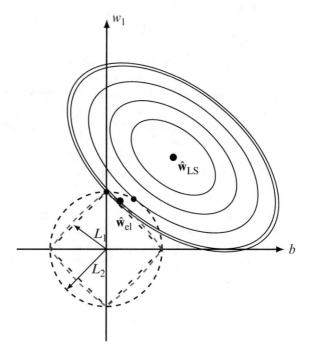

Fig. 2.9 Comparison of elastic net with $\alpha = 0.6$, lasso, and ridge regression results for a two-parameter problem compared with least squares. The curve between the diamond and the circle is the curve $\alpha(|b| + |w_1|) + (1 - \alpha)\sqrt{b^2 + w_1^2} = s$. The resulting solution is in between the ridge and lasso solutions

automatically. For instance if x_1 is multiplied by 0.1 everywhere in the training, the coefficient for the scaled variable will increase by a factor of 10. However, the regularization penalizes the magnitude of the coefficients so that scaling an independent variable may not give the same model. For this reason we typically need to scale the independent and dependent variables so that they are all on a similar, non-dimensional scale. A simple way to normalize is to subtract the mean of the variable, μ_x, and divide by the standard deviation, σ_x:

$$\tilde{x}_i = \frac{x_i - \mu_{x_i}}{\sigma_{x_i}}. \tag{2.31}$$

The mean could be replaced by another measure of the centrality of the variable, such as the median or mode, and the standard deviation could be replaced by the average deviation or some other measure of the range of the variable.

We have not talked about how to determine the penalization strength λ. This is typically handled using cross-validation. A subset of the data is used to produce models with different values of λ. Then the model is tested on the portion of the data not used to train the model. Repeating this several times at each value of λ gives an indication of the model accuracy at each λ value. By varying λ one can see where the ideal trade-off between model accuracy and sparseness/robustness lies. A standard way to choose λ is to pick the largest value of λ that yields a model with mean error within one standard deviation of the minimum error over all λ tested.

Regularization Techniques

The regularization techniques we discussed all modify the loss function so that the model minimizes the sum of the squared error and a penalty term.

- Ridge regression adds a penalty term in the form of the L_2 norm of the coefficients to the loss function to shrink the magnitude of the coefficients.
- Lasso adds a penalty term in the form of the L_1 norm of the coefficients to the loss function. The result, due to the properties of the L_1 norm, will typically set many coefficients to zero.
- Elastic net is a convex combination of the lasso and ridge penalties. It sets groups of parameters to zero and can give models with more predictive power, but the model will not necessarily be sparse.

In all cases the penalty term has a regularization strength parameter that must be chosen using cross-validation. Furthermore, the independent variables should be normalized and non-dimensional so that each affects the penalty in a comparable manner.

2.5 Case Study: Determining Governing Equations from Data

In this case study we will use regularized regression to determine the governing differential equations that generated a data set. In our case we will be dealing with the equations of motion for a pendulum. The pendulum we consider is drawn in Fig. 2.10. The angle between the vertical line crossing the point where the pendulum is mounted and the pendulum is θ. The differential equation governing $\theta(t)$ is

$$\ddot{\theta} + \frac{b}{m}\dot{\theta} + \frac{g}{L}\sin\theta = 0, \tag{2.32}$$

where a dot denotes a time derivative.

A first-order differential equation can be written by defining ω as the angular speed so that

$$\dot{\omega} = -\frac{b}{m}\omega - \frac{g}{L}\sin\theta, \tag{2.33a}$$

$$\dot{\theta} = \omega. \tag{2.33b}$$

Our procedure will be the following. We numerically solve the system of differential equations (2.33) using random initial conditions. The solution will then be known at discrete times, t_i, given as $\theta(t_i)$ and $\omega(t_i) = \dot{\theta}(t_i)$. From the solution we estimate the acceleration of the pendulum using the solution

$$\ddot{\theta} = \dot{\omega} \approx \frac{\omega(t_i) - \omega(t_{i-1})}{t_i - t_{i-1}}.$$

Additionally, we compute a dictionary of functions using the solution at time t_i:

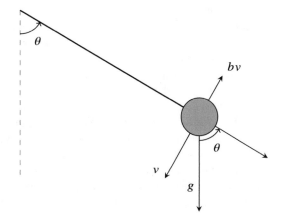

Fig. 2.10 Schematic of the damped pendulum. The acceleration is downward and is due to gravity and the damping force is proportional to the velocity v with proportionality constant b. The angle between the pendulum and the vertical is θ

$$\mathbf{x}(t_i) = \Big\{ \theta, \theta^2, \theta^3, \theta^4, \theta^5, \cos\theta, \sin\theta, \cos^2\theta, \sin^2\theta,$$

$$\omega, \omega^2, \omega^3, \omega^4, \omega^5, \cos\omega, \sin\omega, \cos^2\omega, \sin^2\omega,$$

$$\theta\omega, \theta^2\omega, \theta\omega^2, \theta^3\omega, \theta^2\omega^2, \theta\omega^3, \theta^3\omega^2, \theta^2\omega^3, \sin(\theta\omega), \cos(\theta\omega) \Big\}(t_i).$$

$$(2.34)$$

This dictionary of functions is meant to capture all the possible functions of θ and ω that the acceleration $\ddot{\theta}$ could depend on. This dictionary is not unique and could be much larger or much smaller depending on how much knowledge we have about the potential law of motion.

For a first test, we solve using Runge–Kutta time integration to solve 100 different initial value problems for the pendulum with initial positions θ_0 randomly chosen between $\pm\pi$, with an initial angular speed of $\dot{\theta}_0$ randomly chosen in the interval ± 10 radians per second. From each of the solutions we select 10 random time points to collect the pendulum position, θ, and $\dot{\theta}$ to compute $\ddot{\theta}$ and the dictionary $\mathbf{x}$. In our simulations we set $b = 0.1$ N·s, $m = 0.25$ kg, $g = 9.81$ m/s^2, and $L = 2.5$ m. The collected data are normalized by subtracting the mean and dividing by the norm for each of the elements of $\mathbf{x}$. The penalty parameters for the regularized regression methods are estimated using the cross-validation procedure in the Python library `scikit-learn` [5].

The collected values for $\ddot{\theta}$ versus several of the dictionary elements are shown in Fig. 2.11. The figure also includes the proper linear combination of $\dot{\theta}$ and $\sin\theta$ as well as linear combinations containing the Taylor series of $\sin\theta$. From the figure we can see that there is a strong relationship between several of the dictionary elements that are not the "correct" elements of $\sin\theta$ or $\dot{\theta}$, such as θ, θ^3, and $\theta^2\dot{\theta}$. It is also clear that the correct relationship of $(b/m)\dot{\theta} + (g/L)\sin\theta$ is born out by the data. Using this data we can investigate whether the true relationship can be discovered by regression methods. We use least squares regression and the regularized regression methods lasso, ridge, and elastic net. Using these methods we get different results for the coefficients of the dictionary elements. In Fig. 2.12 the coefficients are compared to the true values of $-b/m = -0.4$ s^{-1} for the $\dot{\theta}$ coefficient, $-g/L = -3.924$ s^{-2} for the $\sin\theta$ coefficient, and 0 for the other coefficients. From the figure we can see that for the $\dot{\theta}$ coefficient all of the methods give values close to the correct value. However, for the coefficient on $\sin\theta$, lasso's result agrees with the true value with the other methods giving a low estimate. Moreover, regularizations using elastic net and ridge give spurious values for the coefficient of θ. This is due to the fact that θ is the leading term in the Taylor series of $\sin\theta$. Therefore, these methods are sharing the contribution to $\ddot{\theta}$ between these inputs. Least squares regression has a spurious non-zero coefficient for θ as well as several others including θ^2 and $\cos\theta$.

While lasso was able to find the proper coefficients in the differential equation from the data, this exercise was a bit too easy in some senses. The only discrepancy we should have expected between the model and the data is numerical error.

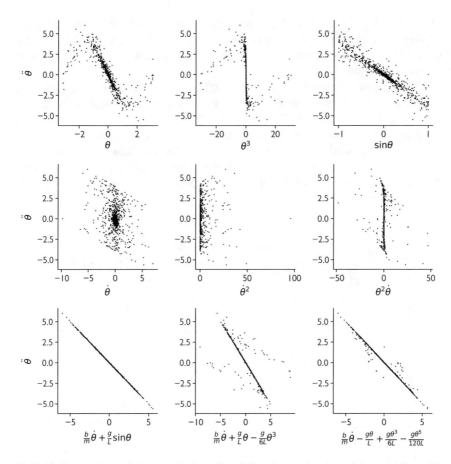

Fig. 2.11 Scatter plots of the numerical results relating the acceleration of the pendulum with several members of the dictionary **x**. The relationship determined by the differential equation is shown in the lower left panel

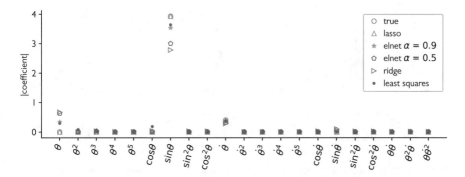

Fig. 2.12 Fit coefficients for a subset of the dictionary elements of **x** using different regression techniques compared with the true differential equation coefficients where only $\dot\theta$ and $\sin\theta$ have non-zero coefficients

Therefore, the acceleration of the pendulum was very nearly exactly what the differential equation predicted. To better test the methods, and to emulate the case where there is measurement error, we repeat the exercise by adding simulated measurement error.

We repeat the exercise above using the same parameters and change the recorded values of θ and $\dot{\theta}$ to each having a random error given by a Gaussian distribution with mean 0 and a standard deviation of 0.01. This in effect gives an unbiased measurement uncertainty to the angle and angular speed measurements of the pendulum. The resulting data, shown in Fig. 2.13, no longer has the precise relationship between $\ddot{\theta}$, $\dot{\theta}$, and $\sin\theta$ that the noiseless case demonstrated. It is clear there is a relationship here, but it is less obvious. Also, it is harder to tell the

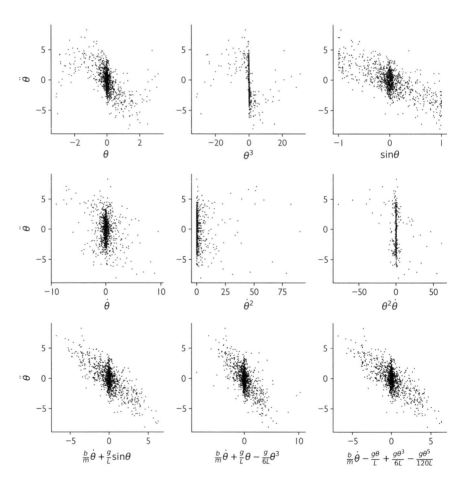

Fig. 2.13 Scatter plots of the numerical results relating the acceleration of the pendulum with several members of the dictionary **x**. The relationship determined by the differential equation is shown in the lower left panel

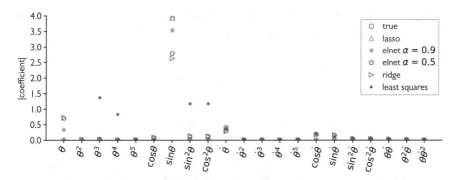

Fig. 2.14 Fit coefficients using the noisy pendulum data for a subset of the dictionary elements of **x** using different regression techniques compared with the true differential equation coefficients where only $\dot{\theta}$ and $\sin\theta$ have non-zero coefficients. Note the least squares coefficients for θ and $\sin\theta$ are not shown because their magnitude is greater than 10

difference between the influence of $\sin\theta$ and its Taylor series approximation due to the presence of noise.

We would expect that the presence of noise would affect the regression estimates for the coefficients. This is most obvious in the results for least squares, as we can see in Fig. 2.14. The least squares results give coefficients of significant magnitude for several dictionary elements, and the coefficients for $\sin\theta$ and θ were greater than 10. The other regression methods also get worse when noise is added, with the exception of lasso. For $\cos\dot{\theta}$ and $\sin\dot{\theta}$ elastic net with $\alpha = 0.5$ and ridge give coefficients significantly different than zero, in addition to the non-zero estimate for $\dot{\theta}$. Once again lasso gives accurate estimates for $\sin\theta$ and $\dot{\theta}$.

2.5.1 Determining the Form of the Coefficients

In the exercise above we used the same pendulum parameters for each of the simulations. We were able to infer the proper coefficients in the differential equation using lasso, but we cannot say how those coefficients will vary as the pendulum mass, length, and damping parameter vary. Given the success of the lasso in determining the coefficients, we will vary the pendulum parameters and repeat the estimation of $\ddot{\theta}$ as a linear combination of $\dot{\theta}$ and $\sin\theta$ using lasso. For each of the different pendula we estimate the coefficients for $\dot{\theta}$ and $\sin\theta$. Using these coefficients we then estimate how they vary as a function of the pendulum parameters.

To posit the variables that the coefficients of $\dot{\theta}$ and $\sin\theta$ depend on we look at units. Knowing that the units of $\ddot{\theta}$ are s^{-1}, we look for combinations of the parameters g (m s^{-2}), L (m), m (kg), and b (kg s^{-1}) that give units of s^{-1} for the coefficient of $\dot{\theta}$, $C_{\dot{\theta}}$, and give s^{-2} for the $\sin\theta$ coefficient, $C_{\sin\theta}$. From this argument we consider the following possible linear combinations:

$$C_{\sin\theta} = w_1\left(\frac{g}{L}\right) + w_2\left(\frac{b^2}{m^2}\right) + w_3\left(\frac{gm}{bL}\right)^2, \qquad (2.35a)$$

$$C_{\dot{\theta}} = w_1\left(\frac{b}{m}\right) + w_2\sqrt{\frac{g}{L}} + w_3\left(\frac{gm}{bL}\right). \qquad (2.35b)$$

The correct linear combination in both cases is $w_1 = -1$ and $w_2 = w_3 = 0$.

To find these coefficients we take 10 different pendula with random values of L, m, and b, and for each we repeat the coefficient estimation exercise and record the coefficients for $\dot{\theta}$ and $\sin\theta$. We also record the values for the functions of b, g, m, and L given in (2.35). From these we can use lasso to estimate the w_i.

From this data collection, we can produce the values in Table 2.1. From these data we estimate the $w_1 = -0.95$ for $C_{\sin\theta}$ and $w_1 = -1.03$ for $C_{\dot{\theta}}$ using lasso regression. The other coefficients are all an order of magnitude smaller. Therefore, we can conclude that our data does indeed return the expected form of the differential equation for $\ddot{\theta}$.

2.5.2   Discussion of the Results

This case study in determining nature's laws from data required more than just machine learning blindly determining the laws based on raw data. For starters, we had to formulate the dictionary of functions that the acceleration could depend on. This step was important because if we did not include the correct functions, our regression procedure may not have found the correct relationships. One of the

Table 2.1 Data collected to estimate the coefficients as a function of the pendulum parameters

$C_{\sin\theta}$	$C_{\dot{\theta}}$	$\frac{g}{L}$	$\frac{b^2}{m^2}$	$\left(\frac{gm}{bL}\right)^2$	$\frac{b}{m}$	$\sqrt{\frac{g}{L}}$	$\frac{gm}{bL}$
−6.795	−0.355	6.722	0.299	150.848	0.547	2.592	12.282
−4.667	−1.142	5.171	1.424	18.771	1.193	2.274	4.332
−3.529	−0.660	3.648	0.544	24.464	0.737	1.910	4.946
−3.762	−2.037	4.340	4.255	4.427	2.063	2.083	2.104
−12.346	−0.283	12.500	0.204	762.995	0.452	3.535	27.622
−19.873	−0.468	19.735	0.329	1183.872	0.573	4.442	34.407
−9.954	−0.492	10.019	0.293	342.102	0.541	3.165	18.496
−3.650	−0.893	3.496	1.010	12.099	1.005	1.869	3.478
−3.653	−0.262	3.611	0.091	142.594	0.302	1.900	11.941
−2.214	−0.413	2.507	0.265	23.709	0.514	1.583	4.869

benefits of the regularized regression techniques is that we can add as many possible dictionary functions as we like.

Additionally, in the step where we determined the form of the coefficients in the differential equation as a function of the pendulum parameters we had to use knowledge of dimensional analysis to propose the possible combinations of parameters. Had we used arbitrary combinations of those parameters it may have been more difficult to capture the correct behavior. Indeed we may have come up with a "law of motion" that made little physical sense.

All of this is to say that scientists and engineers cannot forget their expertise when applying machine learning. Throwing a data set at regularized regression would not have been nearly as fruitful had we not used some scientific knowledge. For example, it would be hard to explain a model with very esoteric units on the coefficients. As it stands, even if we did not know the correct answer for this exercise, we could at least show that the form of the differential equation made sense from a dimensional perspective.

Notes

In our derivations we explicitly formed the matrix $\mathbf{X}^T\mathbf{X}$. The matrix $\mathbf{X}^T\mathbf{X}$ is not usually formed in implementations of least squares models. Rather the singular value decomposition or QR factorization is used to find the weights and biases. This is really an implementation detail in the interest of having an efficient numerical method.

For further discussion of using data to determine physical laws, see the work of Rudy, et al. [6] where several partial differential equations were derived from data, and Brunton, et al. [7] where dynamical systems are considered.

Problems

2.1 Show that scaling an independent variable does not affect the prediction of least squares regression model, but that it does affect the ridge model.

2.2 Consider a data set consisting of 10 samples of three independent variables x_1, x_2, x_3 and one dependent variable y where all three variables are randomly drawn from a standard normal distribution (i.e., a Gaussian with mean 0 and standard deviation 1). Fit three linear least squares models of the form (1) $y = ax_1 + b$, (2) $y = ax_1 + bx_2 + c$, (3) $y = ax_1 + bx_2 + cx_3 + d$. All of these models are nonsensical because they are fitting random noise. Which has a larger value for R^2? Repeat this several times to and comment on your findings.

2.3 Construct a logistic regression model using the following data:

x	0.950	0.572	0.915	0.920	0.520	0.781	0.479	0.461	0.876	0.700
y	1	0	1	1	1	1	0	1	1	0

2.4 Small-angle pendulum

Repeat the exercise of determining the law of motion for a pendulum, but only allow the initial angle to be in the range ± 0.1 radians and having zero initial speed. Use a numerical differential equation integrator of your choice to produce the data.

(a) Derive the approximate law, using Taylor series, for the pendulum in the small-angle limit.
(b) Using the linear different regression techniques estimate the correct differential equation.
(c) What is the effect of noise?

References

1. Michael Gold, Ryan McClarren, and Conor Gaughan. The lessons Oscar taught us: data science and media and entertainment. *Big Data*, 1(2):105–109, 2013.
2. Arthur E Hoerl and Robert W Kennard. Ridge regression: applications to nonorthogonal problems. *Technometrics*, 12(1):69–82, 1970.
3. Robert Tibshirani. Regression shrinkage and selection via the lasso. *Journal of the Royal Statistical Society. Series B (Methodological)*, pages 267–288, 1996.
4. Hui Zou and Trevor Hastie. Regularization and variable selection via the elastic net. *Journal of the Royal Statistical Society: Series B (Statistical Methodology)*, 67(2):301–320, April 2005.
5. F. Pedregosa, G. Varoquaux, A. Gramfort, V. Michel, B. Thirion, O. Grisel, M. Blondel, P. Prettenhofer, R. Weiss, V. Dubourg, J. Vanderplas, A. Passos, D. Cournapeau, M. Brucher, M. Perrot, and E. Duchesnay. Scikit-learn: Machine learning in Python. *Journal of Machine Learning Research*, 12:2825–2830, 2011.
6. Samuel H Rudy, Steven L Brunton, Joshua L Proctor, and J Nathan Kutz. Data-driven discovery of partial differential equations. *Science Advances*, 3(4):e1602614, April 2017.
7. Steven L Brunton, Joshua L Proctor, and J Nathan Kutz. Discovering governing equations from data by sparse identification of nonlinear dynamical systems. *Proceedings of the National Academy of Sciences*, 113(15):3932–3937, April 2016.

Chapter 3
Decision Trees and Random Forests for Regression and Classification

> Mr. Fox: *It's not exactly an evergreen, is it? Aren't there any*
> *pines on the market this side of the river?*
> Weasel: *Pines are pretty hard to come by in your price range.*
>
> —*from the film* Fantastic Mr. Fox

Abstract This chapter covers the topics of decision tree models and random forests. We begin with a discussion of how binary yes/no decisions can be used to build a model for a regression problem by dividing, or partitioning, the independent variables for a simple problem with 2 independent variables. This idea is then generalized for regression problems of more variables before continuing on to classification problems. We then introduce the idea of combining many trees together in the random forest model. Random forests have each tree in the forest built on different training data and use randomization when building the trees as well. The prediction of each of these trees is averaged in a regression problem or votes on a class for a classification problem to give the predicted value of the dependent variable. We demonstrate the efficacy of random forests through a case study of predicting the output of an expensive computer simulation.

Keywords Decision trees · Random forests · Classification trees · Predicting simulations · Calibration

3.1 Decision Trees for Regression

In daily life we often encounter a series of nested binary (i.e., yes/no) decisions. If the sign says "Don't Walk," I do not cross the street. If the weather forecast calls for rain, I bring an umbrella when I leave the house, and if it looks clear later, I may leave the umbrella in my office when I leave for lunch. Decision trees are a way of formalizing these decisions so that they can be cast as machine learning problems.

Fig. 3.1 Visualization of the data in Eq. (3.1). The area of the marker is proportional to the value of y

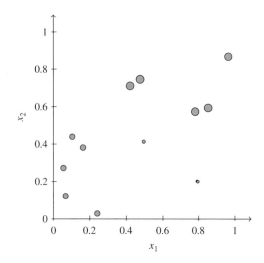

We begin with the regression problem of estimating a dependent variable, y, based on a set of J independent variables $\mathbf{x} = (x_1, \ldots, x_J)$. We seek to partition the independent variables into a number of regions and then assign a value to each of those regions. To demonstrate this consider the independent and dependent variables given by $\mathbf{x} = (x_1, x_2)$ and y

$$
\begin{array}{c|ccccccccccccc}
y & 1.01 & 0.97 & 0.97 & 1.03 & 0.92 & 1.97 & 1.91 & 2.09 & 1.93 & 1.94 & 0.22 & 0.26 & 0.35 \\
\hline
x_1 & 0.06 & 0.10 & 0.24 & 0.16 & 0.07 & 0.42 & 0.96 & 0.48 & 0.85 & 0.78 & 0.79 & 0.80 & 0.50 \\
x_2 & 0.27 & 0.44 & 0.03 & 0.38 & 0.12 & 0.71 & 0.87 & 0.75 & 0.59 & 0.57 & 0.20 & 0.20 & 0.41
\end{array}
\qquad (3.1)
$$

This data set is visualized in Fig. 3.1. We would like to determine a way to split the data orthogonal to an axis to divide the data into two regions. In each of these regions we then assign all of the points a value c that minimizes the squared difference between the true solution and the value c.

In this data set we will split the data using $x_2 = 0.505$ and set the value for all of the points above the split to $\hat{y} = 1.968$; below the split we set all values to $\hat{y} = 0.7162$. The values of 1.968 and 0.7162 are the mean values of the training data that fall into the respective boxes. This splitting of the independent variable space is shown in Fig. 3.2. In the figure we graphically split the independent variable space using a horizontal line at $x_2 = 0.505$; we also display the model as a decision tree. At this point there is only one decision; if x_2 is greater than 0.505 we follow the "yes" path of the tree to arrive at $\hat{y} = 1.968$; otherwise we follow the "no" path to get $\hat{y} = 0.7162$. If we use a squared-error loss function, the loss for this simple tree model is given by

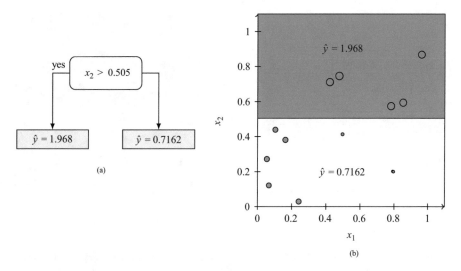

Fig. 3.2 Visualization of the first decision boundary as a tree and in the independent variable space. (**a**) Two-leaf decision tree for predicting $y = f(x_1, x_2)$. (**b**) Visualization of the decision boundary at $x_2 = 0.505$

$$L = \sum_{x_{i2} > 0.505} (y_i - 1.968)^2 + \sum_{x_{i2} < 0.505} (y_i - 0.7162)^2 = 0.9641.$$

Looking at Fig. 3.2b, we see that it may be possible to improve the decision tree: below the split it seems that points toward the right side have smaller values than those to the left. We therefore draw a vertical line downward from the horizontal line at $x_2 = 0.505$ at $x_1 = 0.37$.

Drawing this extra division in the independent variable space gives us two more "leaves" on the tree (that is endpoints in the decision tree). The values of these leaves are $\hat{y} = 0.2767$ when $x_1 > 0.37$ and $x_2 < 0.505$, and $\hat{y} = 0.98$ when $x_1 < 0.37$ and $x_2 < 0.505$. Once again the values of $\hat{y}$ assigned are the average values of the training data inside the boxes. The loss function for this tree model is given by

$$L = \sum_{\mathbf{x}_i \in b_1} (y_i - 1.968)^2 + \sum_{\mathbf{x}_i \in b_2} (y_i - .2767)^2 + \sum_{\mathbf{x}_i \in b_3} (y_i - 0.98)^2 = 0.0365,$$

where $\mathbf{x}_i \in b_\ell$ indicates the values of $\mathbf{x}_i$ leading to the prediction at leaf ℓ in the tree, i.e., $\mathbf{x}_i$ is the box corresponding to leaf ℓ. For example, if $x_2 > 0.505$, the point is in box b_1.

By adding the additional split to the independent variable space, and, as a result, adding another leaf to the tree, we have decreased the loss function by a factor of about 26. This model is much better at explaining the data. We could continue to add splits/leaves until the error was zero by isolating each data point in a box, but this would lead to a very complicated tree model that had nearly as many decisions

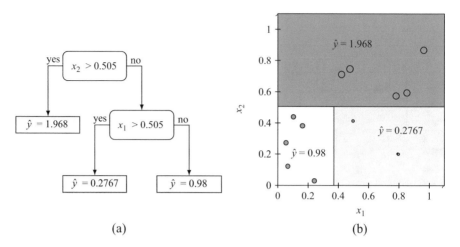

(a)   (b)

Fig. 3.3 Visualization of the two decision boundaries as a tree and in the independent variable space. (**a**) Three-leaf decision tree for predicting $y = f(x_1, x_2)$. (**b**) Visualization of the decision boundaries

in the tree as there are data points. In this example, it appears that adding another split beyond the two we have will not appreciably improve the model (Fig. 3.3).

With this decision tree we can make a few observations. Given the way that the decision boundaries work, all of the boundaries are orthogonal to an axis, and they create rectangular boxes. Using these boxes, if the independent variables fall into a given box, we have a single value for the predicted dependent variable based on the mean of the training data inside that box. Furthermore, we can determine which box a given value of **x** falls into by answering a series of binary yes/no questions.

The decision tree we have created will handle extrapolation in a somewhat graceful manner. If we imagine the decision boundaries extend to infinity when they do not intersect another decision boundary, then any value of **x** will lead to a leaf on the decision tree or a box in the independent variable space. Each of the leaves on the decision tree is the mean of the dependent variables in some subset of the training data. Therefore, the tree model will always give a prediction that is within the range of the training data. This contrasts with the linear models we discussed in the previous chapter where, if an independent variable has a positive coefficient in the model, extrapolating with that variable can lead to a prediction that is arbitrarily large. This is not to say that tree models give the correct answer for extrapolation, but rather their predictions are limited by the training data, and this can be a desirable feature for certain model uses.

We have not yet discussed how we chose where to the place the decision boundaries. Notice that if we were to shift the decision boundaries slightly, the predictions of the decision tree, and the loss function value, would not change. We placed the decision boundaries so that they were equidistant between the nearest training points. This reflects our belief that if multiple places to put the decision boundary give the same loss, we should place the decision boundaries as far from the training data as we can so that a small perturbation to the training data does not give a large change to the prediction.

If we were to generalize the idea of a decision tree to more than two independent variables, the decision boundaries would be a plane in three dimensions, and a hyperplane in higher dimensions. The decision boundaries would still be orthogonal to an axis, and all of the decisions in the tree would still be binary. The tree itself would still have the same appearance as the decision trees above, but visualizing the partitioning of independent variable space becomes much more difficult.

Tree Models
Tree models split the space of independent variables into rectangular regions that each correspond to a different predicted value for the dependent variable. This splitting of the independent variable space is equivalent to a series of yes/no decisions that can be visualized as a decision tree.

3.1.1   Building Regression Trees

We now turn to the mathematical and algorithmic details of building regression trees. We consider a problem with I realizations of a single dependent variable y_i, and J independent variables $\mathbf{x}_i = (x_{i1}, \ldots, x_{iJ})$. We call this data set $\mathbf{X}_0$ and $\mathbf{y}_0$, with "0" to denote that it is the original data set. To build a tree we use a greedy algorithm[1] where we search for a split point s and a splitting variable index j that minimizes the squared-error loss as

$$\min_{s,j} \left[\sum_{x_{ij} \leq s} (y_i - c_-)^2 + \sum_{x_{ij} > s} (y_i - c_+)^2 \right], \tag{3.2}$$

where

$$c_- = \frac{1}{N_-} \sum_{x_{ij} \leq s} y_i, \qquad c_+ = \frac{1}{N_+} \sum_{x_{ij} > s} y_i, \tag{3.3}$$

---

[1] A greedy algorithm searches for the best possible immediate outcome given the known data. An example greedy algorithm would be the hill-climber algorithm in optimization where one always moves in the direction of the gradient of the objective function (i.e., climbs upward on the hill). A greedy algorithm can get stuck at local extreme values, rather than a global maximum/minimum because it typically does not explore enough of the solution space. In the hill-climber algorithm, the solution it finds will be the top of the nearest hill, not the top of the tallest hill. We can use a greedy algorithm to build decision trees because of the pruning step we do later.

and N_+ and N_- are the number of points above and below (or at) the split, respectively. The values of c_- and c_+ are the average values of the dependent variables on either side of the split. To find this split and variable we search through the range of each independent variable to find the split that minimizes the loss.

Once we find this split, we now have two data sets, $(\mathbf{X}_1^-, \mathbf{y}_1^-)$ for all of the cases below the first split and $(\mathbf{X}_1^+, \mathbf{y}_1^+)$ for the cases above. For each of these we can find a best split and variable to split on. That is, we repeat the procedure on the split data sets; we continue in this manner until some stopping criteria are met. A few possible stopping criteria are that we split until:

- There are a certain number of cases remaining in the partitioned data set.
- The cases in the set to be partitioned all have a value within some tolerance.
- The split does not improve the loss function sufficiently.

The first two of these procedures are used in practice to "grow" decision trees. The third option, only splits if it gives enough improvement in the loss function, is not used in practice because it is possible that a split that only provides marginal improvement at one level could lead to a more important partition later on.

Pruning the Tree

The algorithm for growing the tree, using recursive partitions of the data set looking for improvement in the loss function, can grow trees that are very large. Additionally, because we do not account for how much improvement each partition of the data set gives, we do not know what decisions in the tree are responsible for the performance of the tree. Also, because the tree is large with many decision paths, it is possible that the tree will be vulnerable to overfitting (i.e., because the decision tree model is not parsimonious, the likelihood that it is overfitting the training data is increased).

To deal with the size of the tree, we extend the arboreal metaphor and "prune" the trees to remove unnecessary branches/decisions in the tree. We do this by regularizing the loss function in a similar manner to regularized linear regression by adding a penalty to the loss function.

Let T be a decision tree, with T_0 being the tree fit from the procedure outlined above. We then write the number of leaves in the tree, that is the number of places where the tree ends and there are no more decisions to be made, as $|T|$. The regularized loss function for a given decision tree, T, is written as

$$L(T) = \sum_{i=1}^{I} (c_T(\mathbf{x}_i) - y_i)^2 + \lambda |T|, \tag{3.4}$$

where $c_T(\mathbf{x}_i)$ is the predicted value of the dependent variable for tree T given the independent variables for case i, $\mathbf{x}_i$. In this loss function, λ is again a parameter that balances the accuracy of the tree in the first term with the complexity of the tree. As in regularized regression, λ can be chosen using cross-validation.

Using this loss function we search all of the subtrees in T_0 to find the one that minimizes the loss function. We can produce subtrees by removing a split that leads to a leaf. This fuses two leaves together and we can then assess the loss function in Eq. (3.4). To choose the leaves we fuse, we find the leaf that produces the smallest increase in the mean-squared error, the first term in the loss function, when it is removed. We repeat this procedure, fusing leaves and assessing the loss function until we reach the original split in the tree. Comparing the loss function for all of the trees produced will lead to the minimum value of $L(T)$.

> **Regression Trees**
> The procedure for building trees is recursive because each split creates new regions in independent variable space. We then look for partitions in these new regions and repeat until we reach the stopping criteria. However, the tree should be pruned to avoid overfitting through a regularization process where we look for a tree that balances the number of leaves with the error in the predictions.

3.2 Classification Trees

To apply decision trees to classification problems we need to change the way we decide how to split. We could use the cross-entropy loss as we used in regression problems; however, it is more common in building decision trees to use the Gini index. We consider a K-class dependent variable $y \in \{y_1, y_2, \ldots, y_K\}$ and J independent variables $\mathbf{x} = (x_1, x_2, \ldots, x_J)$. The Gini index is a measure of the different classes in a region. In a classification tree consider a partition of the independent variable space that leads to a leaf b_ℓ. In the region of independent variable space we can compute the fraction of the cases in this region that are a member of class k as

$$
p_{\ell k} = \frac{1}{N_\ell} \sum_{\mathbf{x}_i \in b_\ell} I(y_i = k),
$$

where N_ℓ is the number of cases in region b_ℓ and the indicator function $I(y_i = k)$ is given by

$$
I(y_i = k) = \begin{cases} 1 & y_i = k \\ 0 & y_i \neq k \end{cases}.
$$

The indicator function evaluates to 1 for cases where $y_i = k$ and 0 otherwise. To define the Gini index we sum the products of the fraction of each class as

$$G_\ell = \sum_{k=1}^{K} \sum_{k' \neq k} p_{\ell k} \, p_{\ell k'}. \qquad (3.5)$$

The Gini index will be higher when there are many different classes in region ℓ and it will be zero if all of the cases are of the same class. If we consider a two-class problem, $K = 2$, the Gini index simplifies to

$$G_\ell = 2p_{\ell 1}(1 - p_{\ell 1}), \qquad \text{(Gini Index } K=2)$$

because in this case $p_{\ell 2} = 1 - p_{\ell 1}$.

We use the Gini index to decide where to split. Considering a region in independent variable space, we seek a split such that the sum of the Gini indices of the resulting two regions is minimized. The Gini index is a good indicator of where to split because when a split results in a partition with all of the cases on one side having the same class, and the Gini index for that region will be zero.

The procedure for building a classification tree is nearly identical to that for regression: we recursively look for splits that minimize the Gini index of the resulting partition, continuing until a stopping criterion is met. The stopping criteria are also similar. We grow the tree until either they are too few cases to continue or the Gini index is below some threshold. Once the tree is built, we once again need to prune the tree. To prune the tree we use the misclassification rate and the tree complexity. The misclassification rate is given as the proportion of all the cases in a leaf that are not the most common class found in that leaf. For leaf ℓ we call $\hat{k}$ the class that is most common. The misclassification rate, E, is then given by

$$E = \frac{1}{N_\ell} \sum_{\mathbf{x}_i \in b_\ell} I(y_i = \hat{k}).$$

Using the misclassification rate we define a regularized loss,

$$L(T) = \frac{1}{N_\ell} \sum_{\mathbf{x}_i \in b_\ell} I(y_i = \hat{k}) + \lambda |T|, \qquad (3.6)$$

where once again $|T|$ is the number of leaves in the tree. As with regression, to prune the tree we find the subtree of the original tree that minimizes this regularized loss.

Once we have our final, pruned tree, we would like to use it to make predictions. For a given leaf, we can simply give the predicted class as the class that is most common at that leaf. Additionally, we could quote the confidence in the prediction by reporting the misclassification rate or the Gini index for the leaf.

In categorical prediction problems the confusion matrix, $\mathbf{C}$, can be a useful device for reporting the behavior of the model. For a K-class problem, it is a $K \times K$ matrix. The entries C_{mn} are defined as the number of cases in a data set (training or test)

Table 3.1 The confusion
matrix for a hypothetical
model to predict the weather

True/Predicted	Clear	Rain	Snow
Clear	67	12	6
Rain	15	28	3
Snow	18	2	21

that were actually of class m and predicted to be class n by the model, i.e.,

$$C_{mn} = \sum_{i=1}^{I} I(y_i = m)I(\hat{y}_i = n), \tag{3.7}$$

and $\hat{y}_i$ is the predicted class for case i. A perfect model will have a diagonal confusion matrix because all of its predictions will match the true value. The off-diagonals indicate in which way the model is inaccurate. In Table 3.1 a confusion matrix is shown for a hypothetical model that predicts the weather conditions. Looking at this table we see that there were 67 cases where the weather was clear and the model predicted correctly. There were also 12 cases where the weather was clear, but the model predicted rain; 6 times the weather was clear and the model predicted snow. The confusion matrix also shows us that the model is more likely to be wrong by predicting a clear day when the actual conditions were rain or snow ($15 + 18 = 33$) than predicting rain or snow when the conditions were clear. We can also infer from the confusion matrix that the model is not very good at predicting snow: it predicted snow correctly 21 times, and 20 times true conditions of snow were predicted to be rain or clear.

Confusion Matrix
The confusion matrix is a generally applicable technique to understand how classification models perform. It is a matrix of size $K \times K$ for a K-class classification problem. Each entry in the matrix C_{mn} is the number of times that the model predicted class m when the true class was class n. It allows one to assess which classes are accurately predicted or find blind spots in the model by looking at the magnitude of the off-diagonal terms relative to the diagonal values in the matrix. The confusion matrix is the classification analog to a plot of the predicted values versus actual values from a regression problem. The confusion matrix can be computed separately for training and test data to look for overfitting.

The confusion matrix discussion brings up the concept of asymmetric loss. It is possible that from a user's perspective it is much worse if the true conditions are rain and the model predicted a clear day. This would be the case if one were planning a picnic, for example. As another example, in a two-class classification problem, this asymmetry could be that a false negative is much worse than a false positive.

If the consequences of an incorrect prediction differ by case, we want that to be incorporated in our model. This can be done through a risk matrix.[2] The risk matrix **R**, for $K > 2$, is a matrix whose components R_{mn} give the relative consequences of mis-classifying class m as class n. If certain wrong predictions have a worse effect than others, we can reflect this in **R**. The loss matrix is then used to weight the Gini index as

$$G_\ell^{\mathbf{R}} = \sum_{k=1}^{K} \sum_{k' \neq k} R_{kk'} \, p_{\ell k} \, p_{\ell k'}. \tag{3.8}$$

Modifying the Gini index this way in the construction of the tree will try to find splits that lead to minimizing incorrect predictions with higher corresponding values in the risk matrix.

The risk matrix will not work for a two-class problem, $K = 2$, because the Gini index will just account for the sum of the risk matrix elements (this is demonstrated in an exercise). To account for asymmetric loss in a two-class problem, we need to use a different loss function than the Gini index in the tree construction. We can use the misclassification rate in the training and weight that with a risk.

Classification Trees
Classification trees are built in a similar way to regression trees; we partition the independent variable space to find regions where the training data has the same class. The trees are then pruned via a regularization process.

3.3   Random Forests

It is possible to combine many trees together to get a better model of the input–output relationship between independent and dependent variables in a technique known as random forests. The overall idea is that if we can build many small trees that have errors that are not correlated and we average those trees together, we can get a better prediction. We can indulge in a bit of analogizing here: trees grow in forests rather than having one giant tree. In this way we want smaller trees to combine to get a better model than one very large tree.

In random forests we use an extension of the concept of *bagging*. Bagging builds multiple models using different training data: the available data is sampled with replacement (this is known as bootstrapped sampling) to produce a suite of training

---

[2]This is also called a loss matrix, but to avoid confusion with the loss function we use the term *risk* matrix.

sets that models are built on. Because the models are built using different training data, their predictions will be different and can be averaged together.

Random forests [1] modify the concept of bagging to include random sampling of the independent variables. To grow a tree for random forests we produce a bootstrapped sample of the training data. Then using this data we choose where to create splits as in a normal tree, except we only consider a random subset of the independent variables to split on. Using only a subset of the independent variables, coupled with building each tree on a different training set makes the trees behave differently due to different training data and different inputs used to the tree growth. Because the training data and independent variables for the splits are chosen randomly, the trees will have a low correlation with each other so that when we average them together we can have errors cancel, and we get a better prediction.

> **Bagging**
> Bagging is a way to produce different models by modifying the training data that each model uses. We take our original training data and sample with replacement. That is, we pick a random case to put in a training set, and repeat this process until we have the desired number of sampled cases, allowing the cases previously selected to be selected again. Each training set that we produce from this sampling with replacement is then used to build a different model. These models will give different predictions because they were built on different training data.
>
> Random forests modify the concept of bagging to have each tree to select from a random subset of the independent variables when choosing where to split the data.

The algorithm for producing a random forest model with B trees is given below in Algorithm 1. This will produce a set of B trees, T_b. Each of these trees will give a predicted value for the dependent variable $T_b(\mathbf{x})$. We then produce an estimate for the dependent variable $\hat{y}$ as a simple average for regression,

$$\hat{y}(\mathbf{x}) = \frac{1}{B} \sum_{b=1}^{B} T_b(\mathbf{x}), \tag{3.9}$$

or via voting for classification where the class predicted by the most trees is the predicted class for the forest. The value of m is the number of independent variables to consider for a split. The originators of the random forest model suggest $\sqrt{J}$ rounded down to the nearest integer for classification problems and $J/3$, rounded down, for regression problems.

Algorithm 1 Random forest algorithm for producing B trees

---

for $b = 1$ to B **do**
   {Build tree T_b}
   1. Sample with replacement N cases from the training data
   2. Grow a tree by minimizing the appropriate loss function until stopping criteria is met.
   When determining which independent variables to split on, only consider a $m < J$ randomly
   chosen independent variables of the J total variables.
end for

Random Forests

When we combine many decision trees together, we get a forest. The random forest model combines trees but adds randomization two ways:

1. Each decision tree is built using a different sample of the training data.
2. When choosing how to partition the data, only a random subset of the independent variables are considered for each split.

This added randomization makes the trees give different results that, when averaged together in a regression problem or when they vote on a class, are superior to a single large decision tree.

Additionally, due to the way that the random forest is built, we can perform a type of cross-validation while the model is built. Because each tree in the forest sees a different subset of the training data, we can estimate the test error for each tree by using it to evaluate the cases that were not used to train it. Therefore, for each tree we have a prediction at each case not used to train it. These predictions are the out-of-bag estimates (because they were not in the "bag" of training samples). We can combine the out-of-bag estimates from each tree in the forest to get a prediction for each case using the average of the out-of-bag estimates in the regression case, or the majority vote in the classification case. The error in these out-of-bag estimates is an estimate of the test error from cross-validation and can be used to determine how many trees should be in a forest, for example.

From the random forest we can estimate the importance of the independent variables in making a prediction. In each tree in the forest we can record the improvement in the loss function due to splits in each variable. Summing up these improvements over the trees gives an estimate of how important that independent variable is to the model. In addition to the variable importance, we could estimate the uncertainty in a prediction from the random forest by looking at the variability in the estimates from each tree. For a regression problem this could be the standard deviation or variance in the estimate, and for classification we could use the Gini index. Such measures can be useful when one wants to identify important regions in input space where the model is less certain.

One of the benefits that random forests provide over a single tree is the ability to naturally include partitions of independent variable space that are not necessarily orthogonal to the axes. This feature arises from the fact that different trees in the forest will likely put splits in different places on the same variable because of the different cases used to train the trees. Therefore, when we average these slightly different splits, the effective partition of independent variable space will not be orthogonal, which will be seen in an example below.

Furthermore, due to the bagging of the training data, on a regression problem the estimate for the value assigned to a region of independent variable space (i.e., the value at a leaf of the tree) will vary between trees because each tree will have a different mean for the values at a node. This allows random forests to capture the numerical change in the dependent variable without adding too many splits to the tree.

Analysis of Random Forest Models

Because of the way that random forest models are built, there are several beneficial consequences:

- The out-of-bag error behaves in a similar manner to cross-validation error because it is a measure of the error on data that the individual trees have not seen.
- The importance of an independent variable can be estimated based on how many splits in the forest it is involved in.
- The uncertainty in the forest's prediction can be estimated from the variability in the predictions from the individual trees.
- The splits in the data over an ensemble of trees allow for effectively non-orthogonal splits for both regression and classification problems and greater variety in the numeric values predicted for regression problems.

3.3.1 Comparison of Random Forests and Tree Models

To see how random forests behave, and to compare the method to a single tree, we attempt to fit the function

$$y = \begin{cases} 10 & (x_1 - 3)^2 + (x_2 + 3)^2 < 4^2 \\ -10 & (x_1 - 2)^2 + (x_2 - 2)^2 < 1 \\ 2x_1^2 + x_2^2 & \text{otherwise} \end{cases} \qquad (3.10)$$

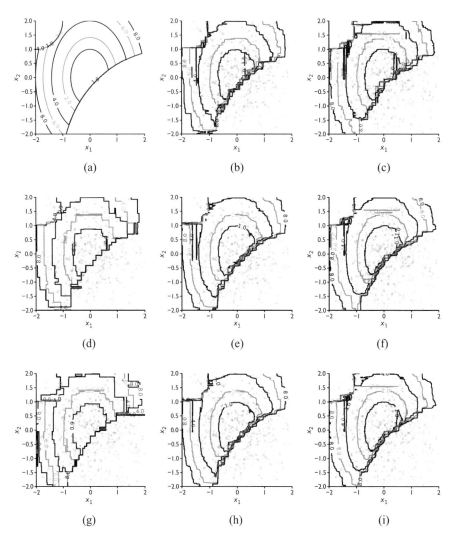

Fig. 3.4 Comparison of contours of $y(x_1, x_2)$ for random forest models with different numbers of trees and a single tree model on two training sets of 2000 points. The numbers in parentheses denote the training set (1 or 2). The training points are also shown. (**a**) True function contours. (**b**) 10 tree forest (1). (**c**) 10 tree forest (2). (**d**) Tree model (1). (**e**) 100 tree forest (1). (**f**) 100 tree forest (2). (**g**) Tree model (2). (**h**) 1000 tree forest (1). (**i**) 1000 tree forest (2)

using training data that is generated by randomly sampling values of x_1 and x_2 from a normal distribution with mean 0 and standard deviation 1.25. Near the origin the function has a smoothly varying region adjacent to two discontinuities leading to a constant value in two circular regions.

In Fig. 3.4 we show the results for fitting a variety of models using two different sets of 2000 training points. The true contour plot of the value of y as functions of

x_1 and x_2 is shown in the top left with contour lines at $y = 1, 2, 4, 6$, and 8. In the first column is the result from a tree model using each of the two training sets. With this large number of training points the differences between the two contour plots are minor. The most noticeable differences are in the regions where there are few training points at the top right and lower left of the plot.

The results from random forest models with different numbers of trees are shown in columns two and three of Fig. 3.4. We have allowed the trees to fit until there was a single case at each leaf when growing the trees and at each split both independent variables were used (rather than a random subset). The rows correspond to different numbers of trees in the forest: 10 trees in row one, 100 trees in row two, and 1000 trees in row three. Between the 10 and 100 tree forests the primary difference is in the estimation of the boundary between the circular region in the lower right and the region where y varies linearly. This transition is smoother in the 100 tree case, although both results are smoother than the single tree model. Additionally, the change in the results for the two different training data sets is more pronounced in the 10 tree model.

The differences between the 100 and 1000 tree forests seem minor when looking at these contour plots. The exact location of the contours varies but it seems to the eye that the contours are in general in agreement with the true contours in the top left of the figure. In the 100 and 1000 tree forests there are some artifacts of the production of the contour plot. There are points where different trees are affecting the value predicted by the model slightly and these artifacts give very narrow contours in certain regions of the problem. These tend to occur in places where the amount of data is small, such as the left side of the plot. The exact location of these regions does seem to depend on the training data.

If we retrain the models using only 200 training points, we can exaggerate the differences between models. The results for this less accurate training are shown in Fig. 3.5. In this figure we see that the differences in the tree model between the two training sets are much more pronounced, and that the tree model produces contours that appear to be rectangular, rather than the rounded shapes of the true function's contours. Additionally, we see that 10 tree random forest model now demonstrates more differences in the results for the two training sets. The random forest models do not estimate as smooth of a transition to the circular region in the lower right as when they were trained with more data. We also note that for a particular data set the difference between the 100 and 1000 tree forests remains minor, as we found in the case with more training points.

To further investigate the behavior of the random forest model we repeat the above exercise using different numbers of trees in the forest: we vary the number of trees from 1 to 200. For each forest we compute the out-of-bag (OOB) error as the variance in the out-of-bag errors for each tree divided by the variance in the dependent variable in the out-of-bag samples. A perfect score for this metric is zero because in this case the errors in the OOB predictions are all zero. This error metric is sometimes called the fraction of variance unexplained because it deals with the variance in the data that cannot be adequately captured in the model.

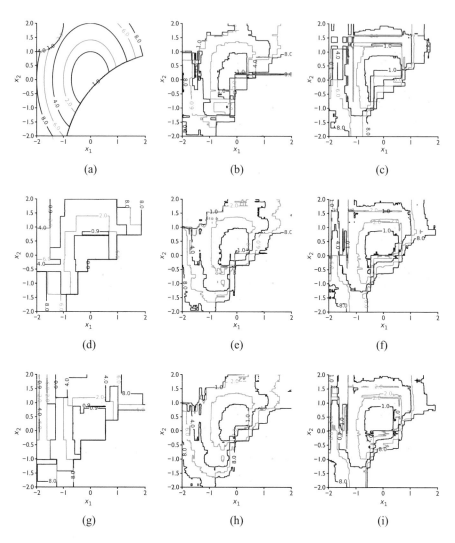

Fig. 3.5 Comparison of contours of $y(x_1, x_2)$ for random forest models with different numbers of trees and a single tree model on two training sets of 200 points. The numbers in parentheses denote the training set (1 or 2). The training points are also shown. (**a**) True function contours. (**b**) 10 tree forest (1). (**c**) 10 tree forest (2). (**d**) Tree model (1). (**e**) 100 tree forest (1). (**f**) 100 tree forest (2). (**g**) Tree model (2). (**h**) 1000 tree forest (1). (**i**) 1000 tree forest (2)

The OOB error for random forest models on this problem as a function of the number of trees in the forest is shown in Fig. 3.6. From the figure we can see that there is clearly a difference in the error in the models based on the number of training points: the error level the models can reach is about an order of magnitude smaller in the case with 2000 training points. Also, from the figure we can see why the results in Figs. 3.4 and 3.5 demonstrated that the 100 tree forest was different than

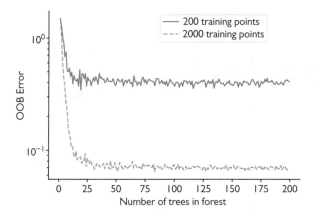

Fig. 3.6 The OOB error given as the fraction of variance unexplained by the model

the 10 forest model, but not noticeably different than the 1000 tree forest. It seems that the OOB error has converged within the variability of the training set variations between 25 and 50 trees in the forest. Beyond this point it appears that adding more trees has little effect, as we saw in the above. In hindsight, we could have used this result to decide how many trees were enough for our data.

3.4 Case Study: Predicting the Result of a Simulation Using Random Forests

As a case study for random forests we will consider the problem of predicting the output of a computational physics simulation code. Many computer simulations are expensive to run in terms of both computational time and human time. The computational time cost is understood by any person who has had to wait for a simulation to complete as part of a larger analysis. The cost of human time, while harder to measure, is also important. Some computer simulations require effort to set up, whether in problem specification or in generating a computational mesh or other setup costs. Other costs include dealing with failed runs due to incorrect setting of parameters in the simulation such as iterative tolerances, time steps, or other model parameters.

The particular problem we will discuss is the computational simulation of a laser-driven shock launched into a beryllium (Be) disc as part of a high-energy density physics experiment [2–4]. An illustration of the experiment is shown in Fig. 3.7 In this problem we use the radiation-hydrodynamics code Hyades [5] to predict when a shock wave created by a high-energy laser will exit the Be disc, that is, when the shock "breaks out" of the Be. The shock is created when the laser ablates some of the Be from the disc. The momentum of this ablated material creates a counter

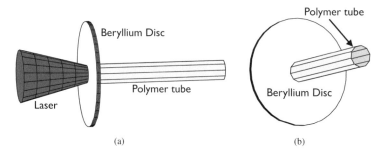

Fig. 3.7 Illustration of the experiment being modeled in the simulations: a laser strikes a beryllium disc, launching a shock wave into the beryllium that eventually leaves the beryllium, i.e., "breaks out" into the gas-filled, polymer tube. The disc thickness is exaggerated in the drawings because it is only about 20 microns thick. (**a**) Side view. (**b**) End view

flowing shock into the disc. This shock eventually reaches the back surface of the disc and breaks out into the gas-filled tube. At this breakout time the conditions of the disc are used to feed another simulation of the shock propagating down a gas-filled polymer tube.

Hyades computes the hydrodynamic motion using a computational mesh that moves with the materials. This means that if there is significant mixing or turbulent flow, the mesh can become twisted to the point of the simulation crashing due to mesh zones being twisted in such a way that the effective volume of a zone is negative. To deal with such mesh twisting sometimes requires hand adjusting of the mesh to remove these tangled zones. This is a source of much human effort, along with the computational cost of the simulations.

To this end one may want to build a machine learning model, often called a surrogate model, to predict what the Hyades output would be, in this case the shock breakout time. For these particular simulations there are 5 parameters that we could consider varying. The Be disc thickness can be varying because the thickness of the disc can vary within several microns (μm) due to machining tolerances. The second parameter is the laser energy because there is shot-to-shot variation in the actual amount of energy that the laser can deliver to the disc. In addition to these two physical parameters, there are three model parameters to vary. These model parameters are often set to match some experimental observation, and as such they are sometimes called tuning parameters. These tuning parameters are used to deal with the fact that the simulation uses some approximate models to compute the system evolution. In the code's mathematical models we use a gamma-law equation of state for Be, and the value of γ in the equation of state is a parameter we can vary. Additionally, the value of the flux limiter for the electron diffusion model is another parameter that can vary, as well as a parameter called the wall opacity that is a multiplier on the optical thickness of the tube wall. These five parameters (2 physical and 3 model parameters) will be our independent variables and the dependent variable will be the shock breakout time.

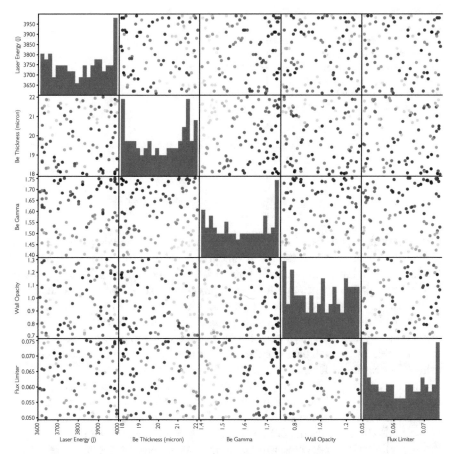

Fig. 3.8 The 104 sets input parameters for the Hyades simulation runs. The diagonal plots show the histogram of the samples for each independent variable, and the off-diagonals show the scatter plot of the independent variables projected onto the 2-D planes of each pair of variables. The scatter plot points are shaded so that points with a shorter breakout time are darker

For these five parameters we have sampled 104 points[3] and run Hyades calculations for each and recorded the shock breakout time. The sampled points projected onto the two-dimensional planes of each pair of coordinates are shown in Fig. 3.8, as well as the histograms for the samples of each independent variable. The samples do a reasonable job of covering the five-dimensional input space. Additionally, this figure shades the points in the scatter plots so that simulations with a shorter breakout time are darker.

---

[3]These are a subset of the original planned points, but several simulations could not be run to completion.

Fig. 3.9 The out-of-bag error, measured as the fraction of variance unexplained, for the random forest model predicting the shock breakout time as a function of the number of trees in the forest

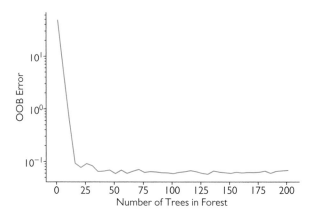

Our task now is to use these simulations to produce a random forest model to predict the shock breakout time as a function of the five inputs. For this we use the implementation of the random forest model in the library scikit-learn for Python [6]. We break our data into training and test sets with 15 simulations in the test set. We then compute the OOB error, which is the fraction of variance unexplained, for random forest models using different numbers of trees. We see in Fig. 3.9 that by 100 trees per forest, the OOB error seems to have converged. Also, given our relatively small data size there is little downside in terms of model cost by using 200 trees in the forest so we use that for our final model.

The next step is to build the model and compare its accuracy on the test and training sets. In Fig. 3.10 we show the predicted values for the shock breakout from the random forest model versus the true values in the data. A perfect model would have all of the results fall on the 45° line $x = y$. From the results we see that the training data falls closer to the correct predictions relative to the test data. However, there does not appear to be any bias or systematic flaws in the predictions for the training points. For the test data the mean-absolute error was 0.013 ns and the mean-absolute percent error was 3.35%. These are both acceptable given the large variation of breakout times in the simulations. The mean-absolute error (MAE) is defined as

$$L_{MAE} = \frac{1}{I} \sum_{i=1}^{I} |\hat{y}_i - y_i|, \tag{3.11}$$

and the mean-absolute percent error (MAPE) is given by

$$L_{MAPE} = \frac{100}{I} \sum_{i=1}^{I} \frac{|\hat{y}_i - y_i|}{|y_i|}. \tag{3.12}$$

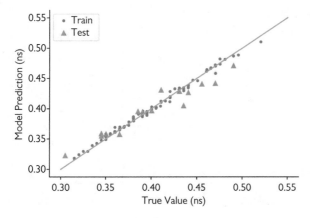

Fig. 3.10 Predicted versus actual shock breakout times (in nanoseconds) for the 200 tree random forest model

Table 3.2 The variable importances provided by `scikit-learn` for the shock breakout random forest model

Variable	Laser energy	Be thickness	Be gamma	Wall opacity	Flux limiter
Importance	0.010	0.188	0.731	0.013	0.058

Now that we have a random forest model that can predict the shock breakout time given the five inputs, what should we do with it? We can use this model to understand what inputs are most important in the simulation. The random forest model in `scikit-learn` does give a variable importance measure based on the decrease in variance attributed to each input. The numbers are reported as a number for each independent variable and the total importances sum to 1. These values are reported in Table 3.2. From these outputs we see that the value used in the gamma-law equation of state, Be Gamma in the table, has the largest importance, followed by the thickness of the Be disc, and the flux limiter value. The laser energy and the wall opacity have seemingly little importance. For the laser energy, this can be explained by the fact that the range of laser energies was not that large and the fact that differences in the laser energy can be counteracted by changing other parameters. The wall opacity not being important is due to the fact that the optical properties of the polymer tube do not influence the shock behavior prior to break out.

To get a better sense of how these parameters can affect the predicted breakout time, we will plot the main effects of the model. The main effect for an independent variable is the average behavior of the output when we average the predictions over all independent variables except for one. In Fig. 3.11 we plot the main effects for each of the independent variables. To compute these main effects we compute the average model prediction at a given value of a single independent variable and using the available data for all of the other independent variables. For example the main effect for variable x_1, the laser energy, would be computed as

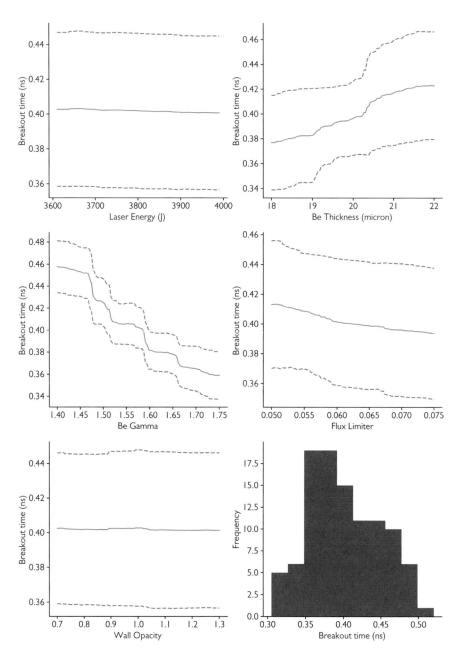

Fig. 3.11 The main effects, as calculated via Eq. (3.13), for the random forest model. The dashed lines are plus/minus one standard deviation of the model outputs as a function of the independent variable. For reference in the lower right the histogram of the breakout times in the simulation data is shown

$$m(x_1) = \frac{1}{I} \sum_{i=1}^{I} f(x_1, x_{i2}, x_{i3}, x_{i4}, x_{i5}), \qquad (3.13)$$

where I is the number of available data points and x_{ij} is the ith value of variable j in the data set. The value $f(x_1, x_{i2}, x_{i3}, x_{i4}, x_{i5})$ is computed using the random forest model. This calculation can be performed for each independent variable. The variables that are not important in the model will have their main effect functions be relatively flat, indicating that varying just this independent variable does not affect the mean output of the model. Also, at each value of the independent variables that we are computing the main effects we can compute a standard deviation of the I model outputs at that point.

For the main effects plotted in Fig. 3.11 we can see that the value of Be gamma and the disc thickness are two important variables. The mean prediction changes noticeably as these variables change, and the standard deviation bands are also narrower. We see that increasing the value of Be gamma decreases the shock breakout time. The main effects also show that increasing the disc thickness increases the breakout time, as one might expect.

The flux limiter parameter demonstrates some influence on the model predictions: the main effects for this variable change between the lowest value and the largest value of the flux limiter. However, the laser energy and the wall opacity have main effects curves that appear flat. This indicates that changing these variables is much less important that the other variables. Moreover, the ± 1 standard deviation band for these variables is nearly the entire range of the breakout data, as can be seen from the histogram. In this case the importances we infer from the main effects are the same as from the reported importances from scikit-learn, but the extra work is worth it because we can get a better understanding of how the model behaves. The cost of computing the main effects is also modest because we are just evaluating the random forest model repeatedly.

3.4.1 Using the Random Forest Model to Calibrate the Simulation

The random forest model can be thought of as an inexpensive way to estimate what a full simulation would calculate the shock breakout time to be. One possible use of this tool is to determine what the values of the simulation parameters should be to get a desired result. For a given experiment there will be a value of the disc thickness and the laser energy, but the value of the Be gamma, the flux limiter, and the wall opacity are free to be modified to match the experimental breakout time. We consider a scenario where we have a disc of thickness $20.16\,\mu\text{m}$, a laser energy of $3820\,\text{J}$, and a measured breakout time of $0.39\,\text{ns}$. We wish to use our random forest model to determine what the correct values of Be gamma, the flux limiter, and the wall opacity are to match this experiment.

To do this we first set the wall opacity to the median value from the previous simulation data, 1.005. This is done because our analysis above showed that it had minimal impact on the shock breakout time. We now need to find the values of Be gamma and the flux limiter that we need to match the simulation. We know that our random forest model is not perfect because there is some error in the predictions. Using the test data we estimate the standard deviation of the error on the test data to be $\sigma = 0.01566$ ns. We then use this standard deviation to score a particular value of a random forest prediction as

$$\text{score}(\mathbf{x}) = \exp\left(-\frac{(f(\mathbf{x}) - 0.39)^2}{\sigma^2}\right). \qquad (3.14)$$

We vary the values of Be gamma and the flux limiter over their respective ranges in the original data while fixing the values of the laser energy, disc thickness, and wall opacity at the values given above. The value of the score will be 1 if the random forest model predicts the desired value of 0.39 ns and the score decreases as the prediction moves away from the desired value for the shock breakout time. The standard deviation scales how fast this decrease occurs. If the model predicts a value that is less than a standard deviation away from the desired value, the score will be higher than if the prediction is farther away. This also implies that our calibration will be better if the standard deviation is small because in that scenario only values quite close to the desired value will get a high score.

Note that our scoring mechanism is symmetric in that predictions above or below the target of the same magnitude would get the same score. We have implicitly assumed that the random forest's errors are symmetric. This is reasonable because in our test we did not observe a strong bias in the predictions. If we had seen this, we might have to adjust the way we score a prediction. Furthermore, if we had evidence that σ was also a function of the inputs, we might have to adjust our scoring as well.

In Fig. 3.12 we plot the score as a function of the two calibration parameters. To create this figure we evaluated the random forest model at a grid of 200×200 points

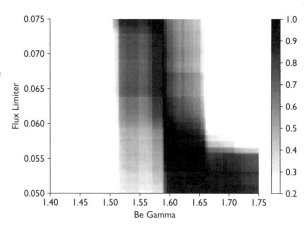

Fig. 3.12 The results from the calibration score as given by Eq. (3.14) for calibrating Be gamma and the flux limiter to give a breakout time of 0.39 ns for a disc of thickness 20.16 μm, a laser energy of 3820 J, and a wall opacity of 1.005

for Be gamma and the flux limiter. In this figure the dark regions correspond to high score regions of input space. The calibration reflects the fact that Be gamma has a stronger influence on shock breakout than the flux limiter: for every value of the flux limiter we can get a high score, but only certain values of Be gamma give a high score.

We can use the calibration to decide, for example, where we should run the full Hyades code to get the desired breakout time. This could represent a large saving in human and computer time because we would not be required to run multiple simulations to perform the calibration. One question that naturally arises is what value of Be gamma and the flux limiter should one use in running the code. In Fig. 3.12 we see that there is a region between Be gamma of about 1.60 and 1.65 and a flux limiter range of 0.0525 and 0.0575 where the score is near one. Given that this plateau of high score is the largest such region, it is sensible to pick a value in the centroid of this region. This selection is preferred to, for example, a point on the ridge of high score just below Be gamma of 1.60. At this value of Be gamma it seems any value of the flux limiter would allow a high score, but a slight change in Be gamma has the score drastically decrease for certain values of the flux limiter.

If multiple runs of the code were possible, one could sample points for Be gamma and the flux limiter using the scores and rejection sampling [7]. One picks at random values for Be gamma and the flux limiter in their respective ranges. Then one picks another random number, ξ, between 0 and 1. If the score at the selected values of Be gamma and the flux limiter is larger than ξ, we then run those values through a full simulation. Otherwise, the point is rejected and we sample the values and ξ again. Doing this repeatedly will pick points that have their density distributed in the same way as the scores in Fig. 3.12.

This calibration exercise is just one way we can use our random forest model. We could perform a calibration for multiple values. Say we have a suite of experiments with different disc thicknesses and laser energies, we could have the score reflect what values of the inputs match multiple experiments by multiplying exponentials together, for instance. Additionally, given that we found in our model that Be gamma was an important parameter in determining the shock breakout time, that indicates that an investment in improving the equation of state model in the simulation (where the Be gamma was a parameter) would have a bigger impact on the ability of the code to model shock breakout time than, for example, improving the electron diffusion calculation because the flux limiter parameter in that model had a much smaller impact.

Notes

Our discussion of tree models did not touch on some other machine learning models that are based on trees. For a detailed overview of these models, and a more mathematically rigorous treatment of random forest models, please see [8]. The models include:

- Multiple adaptive regression splines (MARS) where instead of a constant value at the leaves of the tree, one uses a spline function to represent the variation in the dependent variable.
- AdaBoost where a series of weak trees are created where each set of trees has a different weighting of the training data based on the errors of the previous set of trees.

It is also worth mentioning that random forests have been successful in modeling experiments and enabled a unique calibration procedure that allowed for a new understanding of how nuclear fusion experiments should be conducted [9]. Using a random forest model as a surrogate for expensive simulations, researchers were able to determine that compressing hydrogen symmetrically was not the only way to get fusion in an inertial confinement experiment. Because the random forest model was inexpensive to evaluate, they determined (and subsequent simulations confirmed) that certain asymmetric compressions were more stable to perturbations than symmetric compression.

Finally, random forests have been used to initialize neural networks, a topic we discuss in a subsequent chapter. The trees in the forest can be mapped to a neural network as a way to warm start the network training and somewhat determine the architecture of the network [10].

Problems

3.1 Consider a two-class classification problem and a generic risk matrix **R**. Show that the weighted Gini index given in Eq. (3.8) will not affect the building of a decision tree.

The remaining problems can be completed using a library such as `scikit-learn`.

3.2 Construct a training data set for a regression problem using the formula $y = 10x_1 + x_2 + 0.1x_3 + 0.01x_4 + 0.001x_5$ by sampling each x_i from a standard normal distribution (that is a normal distribution with mean 0 and standard deviation 1). Use 100 cases, that is 100 samples for the x_i and 100 corresponding values for y. Compare the following models: (1) a single decision tree, (2) random forests, and (3) linear regression using a test sample. Repeat the comparison using 10 and 1000 cases in the training and comment on the changes.

3.3 Consider a problem where there are two independent variables z and τ and the dependent variable is $U(z, \tau)$. In Table 3.3 blank entries are zero. Fit the data using a decision tree and random forests.

3.4 Build decision trees and random forests on the Rosenbrock function

$$f(x, y) = (1 - x)^2 + 100(y - x^2)^2,$$

Table 3.3 $U(z, \tau)$

z/τ	0.1	0.316228	1	3.162278	10	31.622777	100
0.01	0.004657	0.038104	0.216108	0.706021	1.447721	0.388874	0.188821
0.1	0.004637	0.037628	0.213206	0.698641	1.437848	0.388742	0.188807
0.17783	0.004577	0.036528	0.206771	0.682453	1.416247	0.388454	0.188778
0.31623	0.004238	0.032486	0.185556	0.630645	1.347603	0.387544	0.188684
0.45	0.003151	0.024726	0.151256	0.551223	1.243751	0.386186	0.188544
0.5	0.002341	0.020397	0.134328	0.513701	1.195232	0.385559	0.188479
0.56234	0.001374	0.015126	0.113312	0.466215	1.132999	0.384686	0.188388
0.75	0.000227	0.005791	0.066626	0.345870	0.963841	0.381462	0.188052
1	0.000012	0.001386	0.031387	0.228898	0.775692	0.375816	0.187456
1.33352		0.000153	0.010575	0.128652	0.578770	0.366022	0.186402
1.77828		0.000005	0.002124	0.057239	0.389652	0.349363	0.184541
3.16228				0.003761	0.111381	0.279376	0.175633
5.62341					0.012613	0.146096	0.150316
10					0.000266	0.026726	0.092592
17.78279						0.000470	0.021382

by sampling 100 points randomly in the interval $[-1, 1]$ to create the training set. Use the models to find the location of the minimum of the function and compare your result with the true values of $x = -1$ and $y = 1$.

3.5 Generate a classification data set by sampling values for x_1 and x_2 randomly in the interval $[0, 1]$. Then evaluate the function

$$f(x_1, x_2) = \exp\left(-\frac{x_1^2 - x_2^2}{10}\right),$$

and compare the values to another random number, ξ, in the interval $[0, 1]$. If $\xi > f(x_1, x_2)$, the case is in class 1; otherwise, the case is in class 0. Build a random forest model and decision tree model for the data you generate. Visualize the decision boundaries and evaluate the confusion matrix for models built with 10, 100, and 1000 cases.

References

1. Leo Breiman. Random forests. *Machine learning*, 45(1):5–32, 2001.
2. Ryan G McClarren, R Paul Drake, J E Morel, and James Paul Holloway. Theory of radiative shocks in the mixed, optically thick-thin case. *Physics of Plasmas*, 17(9):093301, 2010.
3. R P Drake, F W Doss, R G McClarren, M L Adams, N Amato, D Bingham, C C Chou, C DiStefano, K Fidkowski, B Fryxell, T I Gombosi, M J Grosskopf, J P Holloway, B van der Holst, C M Huntington, S Karni, C M Krauland, C C Kuranz, E Larsen, B van Leer, B Mallick, D Marion, W Martin, J E Morel, E S Myra, V Nair, K G Powell, L Rauchwerger,

P Roe, E Rutter, I V Sokolov, Q Stout, B R Torralva, G Toth, K Thornton, and A J Visco. Radiative effects in radiative shocks in shock tubes. *High Energy Density Physics*, 7(3):130–140, September 2011.

4. H F Stripling, R G McClarren, C C Kuranz, M J Grosskopf, E Rutter, and B R Torralva. A calibration and data assimilation method using the Bayesian MARS emulator. *Annals of Nuclear Energy*, 52(C):103–112, February 2013.

5. Jon T Larsen and Stephen M Lane. HYADES—A plasma hydrodynamics code for dense plasma studies. *Journal of Quantitative Spectroscopy and Radiative Transfer*, 51(1-2):179–186, 1994.

6. F. Pedregosa, G. Varoquaux, A. Gramfort, V. Michel, B. Thirion, O. Grisel, M. Blondel, P. Prettenhofer, R. Weiss, V. Dubourg, J. Vanderplas, A. Passos, D. Cournapeau, M. Brucher, M. Perrot, and E. Duchesnay. Scikit-learn: Machine learning in Python. *Journal of Machine Learning Research*, 12:2825–2830, 2011.

7. Ryan G McClarren. *Uncertainty Quantification and Predictive Computational Science*. Springer, 2018.

8. Trevor Hastie, Robert Tibshirani, and Jerome Friedman. *The Elements of Statistical Learning*. Data Mining, Inference, and Prediction, Second Edition. Springer Science and Business Media, New York, NY, August 2009.

9. J L Peterson, K D Humbird, J E Field, S T Brandon, S H Langer, R C Nora, B K Spears, and P T Springer. Zonal flow generation in inertial confinement fusion implosions. *Physics of Plasmas*, 24(3):032702, 2017.

10. Kelli D Humbird, J Luc Peterson, and Ryan G McClarren. Deep neural network initialization with decision trees. *IEEE transactions on neural networks and learning systems*, 2018.

Chapter 4
Finding Structure Within a Data Set: Data Reduction and Clustering

Whose heat doth force us thither to intend,
But soule we finde too earthly to ascend,
'Till slow accesse hath made it wholy pure,
Able immortall clearnesse to endure.

—John Donne from "To the Countesse of Huntington"

Abstract This chapter covers unsupervised learning techniques that attempt to extract structure out of data without being told information about a dependent variable. The idea is that we have a set of data and we want to know if there are patterns that arise in the data. The first method we discuss, the singular value decomposition (SVD), treats the data as a matrix and looks for a structure in that matrix. We apply the SVD to a set of time series data from simulation of high-energy density physics experiments and use the results in a supervised learning problem. We also cover K-means clustering, where clusters in the data set are found using distance measures in the independent variables, and t-SNE, where high-dimensional data are mapped into a low-dimensional (2 or 3 dimensions) data set to visualize the clusters. We close this chapter by applying supervised learning methods to hyperspectral imaging of plant leaves.

Keywords Singular value decomposition (SVD) · Data reduction with SVD · Clustering · k-means algorithm · t-Distributed Stochastic Neighbor Embedding (t-SNE) · Hyperspectral imaging

4.1 Singular Value Decomposition

The previous chapters on linear regression and tree-based models focused on supervised learning problems where we wanted to understand the functional relationship between independent variables and dependent variables, that is, we wanted to find approximations to the relationship $\mathbf{y} = f(\mathbf{x})$. In this chapter we approach the problem of unsupervised learning where we have a set of data comprised of

© Springer Nature Switzerland AG 2021
R. G. McClarren, *Machine Learning for Engineers*,
https://doi.org/10.1007/978-3-030-70388-2_4

many variables and we want to find patterns or relationships between the different variables.

There are many scenarios where one may have several different measurements for each case. These measurements could be the size of different body parts (head, arms, feet, etc.) in a human being, the performance characteristics of an automobile, or the pixel values in an image. These measurements are telling us about the underlying cases, but not all the measurements are independent. For example, in the case of the human body someone with large arms will tend to have large feet, but not always. We would like to take the measurements that we have and find ways to combine them into uncorrelated measurements that tell us what are ways that our data tends to vary.

To do this we consider a data set that is comprised of I cases and that each case has J measurements or variables associated with it. We write each case as $\mathbf{x}_i = (x_{i1}, x_{i2}, \ldots, x_{ij} \ldots, x_{iJ})$ for $i = 1, \ldots, I$. We can assemble these cases into a matrix $\mathbf{X}$ of size $I \times J$ where each row is a case and each column corresponds to a given variable. We are interested in how the columns of $\mathbf{X}$ are related: that is how the measurements may vary together or be correlated. We accomplish this using the singular value decomposition (SVD). For the discussion below we assume that $I > J$.

The SVD of the matrix $\mathbf{X}$ factors $\mathbf{X}$ into three matrices:

$$\mathbf{X} = \mathbf{USV}^\mathrm{T}, \tag{4.1}$$

where U is an $I \times J$ orthogonal matrix,[1] S is a $J \times J$ matrix with zeros everywhere except the diagonal where there are non-negative values, and $\mathbf{V}$ is a $J \times J$ orthogonal matrix.

The diagonal matrix $\mathbf{S}$ is arranged such that the diagonal elements are in decreasing order. We write s_i as the diagonal element for row i and order $s_1 \geq s_2 \geq s_3 \cdots \geq s_J$. We call the s_i singular values of the matrix. It is possible that not all of the singular values are non-zero. For example, if only $r < J$ of the singular values are nonzero, then we can factor $\mathbf{X}$ by dropping the rows in $\mathbf{U}$ and $\mathbf{V}$ that would be multiplied by the zero singular values as

$$\mathbf{X} = \mathbf{U}_r \mathbf{S}_r \mathbf{V}_r^\mathrm{T}, \tag{4.2}$$

where $\mathbf{U}_r$ is an $I \times r$ matrix, $\mathbf{S}_r$ is a diagonal matrix of size $r \times r$, and $\mathbf{V}$ is a matrix of size $J \times r$.

The number of non-zero singular values in the matrix is the rank of the matrix. As a result, the number of non-singular values tells us the number of independent variables that describe the cases in the data. If $r < J$, then we can exactly represent the data with only r variables instead of all J.

---

[1] A matrix is orthogonal if the product of the matrix with its transpose is the identity matrix, $\mathbf{I}$.

The matrix $\mathbf{U}$ has the same number of rows as the original data matrix $\mathbf{X}$. Each column in $\mathbf{U}$ represents the value of a new variable that is uncorrelated with the other columns. Each of these columns is a linear combination of the original columns in the data matrix. This linear combination is given by the columns of the matrix $\mathbf{V}$ (or, equivalently, the rows of $\mathbf{V}^{\mathsf{T}}$) times the corresponding singular value. Given that the singular values are ordered in decreasing magnitude, the columns of $\mathbf{U}$ give the uncorrelated variables in decreasing order of importance. Therefore, we can understand the cases of our data by looking at the first few columns in $\mathbf{U}$ in many instances.

The columns of $\mathbf{V}$ define a linear combination of the original measurement values to form a new uncorrelated variable. We can try to interpret what these mean by looking at the changes as we add terms to the SVD reconstruction of the original matrix. This will indicate what effects the different uncorrelated variables contribute to. In the case study below we use this approach to interpret the uncorrelated variables.

SVD to Approximate a Matrix

Even if all p singular values are non-zero, it is possible that several of the singular values are approximately zero or that the first few singular values are much larger than the others. Consider the fraction m_ℓ defined as

$$m_\ell = \frac{\sum\limits_{n=1}^{\ell} s_n}{\sum\limits_{n=1}^{p} s_n}. \tag{4.3}$$

The value of $m_p = m_r = 1$. The value of m_ℓ is sometimes called the fraction of variation explained. That is, it gives an indication of how well the first ℓ columns of $\mathbf{U}$ can reproduce the data in the original matrix.

Using m_ℓ we can compute how well the first ℓ singular values approximate the variability between cases in the matrix. For instance, say that a particular integer t gives $m_t = 0.9$, and then we know that the first t singular values approximate the 90% of the variability between cases. And as a result there are t combinations of the variables given by the first t columns of $\mathbf{V}$ that explain nearly all of the variability in the data.

The approximate matrix using the first ℓ singular values is the product of smaller matrices

$$\mathbf{X} \approx \mathbf{U}_\ell \mathbf{S}_\ell \mathbf{V}_\ell^{\mathsf{T}}, \tag{4.4}$$

where $\mathbf{U}_r$ is an $I \times \ell$ matrix, $\mathbf{S}_\ell$ is a diagonal matrix of size $\ell \times \ell$, and $\mathbf{V}$ is a matrix of size $J \times \ell$. The resulting matrix is only an approximation to $\mathbf{X}$, and the accuracy of the result depends on how close m_ℓ is to 1.

To illustrate the properties of the SVD we will consider a simple example of a data matrix, $\mathbf{A}$, that has 6 cases each with 4 measurements, that is, a 6×4 matrix. First, we write out the SVD of this matrix:

$$\begin{pmatrix} 0.95 & 0.37 & 0.58 & 0.18 \\ 0.59 & 0.15 & 0.40 & 0.11 \\ 0.29 & 0.22 & 0.11 & 0.03 \\ 0.12 & 0.12 & 0.04 & 0.05 \\ 0.23 & 0.23 & 0.07 & 0.03 \\ 1.24 & 0.51 & 0.74 & 0.23 \end{pmatrix} =$$
$$\mathbf{A}$$

$$\begin{pmatrix} 0.554 & -0.101 & 0.092 & 0.140 \\ 0.341 & -0.443 & -0.118 & -0.011 \\ 0.169 & 0.516 & -0.477 & -0.657 \\ 0.076 & 0.323 & 0.858 & -0.289 \\ 0.139 & 0.650 & -0.114 & 0.680 \\ 0.723 & 0.007 & 0.029 & -0.049 \end{pmatrix} \begin{pmatrix} 2.141 & 0 & 0 & 0 \\ 0 & 0.233 & 0 & 0 \\ 0 & 0 & 0.036 & 0 \\ 0 & 0 & 0 & 0.006 \end{pmatrix} \begin{pmatrix} 0.801 & 0.328 & 0.478 & 0.148 \\ -0.047 & 0.865 & -0.496 & -0.061 \\ -0.229 & 0.075 & 0.033 & 0.970 \\ -0.551 & 0.372 & 0.724 & -0.183 \end{pmatrix} .$$
$$\mathbf{U} \qquad\qquad \mathbf{S} \qquad\qquad \mathbf{V}^{\mathrm{T}}$$

From the matrix $\mathbf{S}$ we see that the first element in the diagonal is much larger than the other entries. This indicates that the columns are not independent measurements and that there is some dependence between them. We can further see this by looking at the values of m_ℓ: $m_1 = 0.8861$, $m_2 = 0.9825$, and $m_3 = 0.9976$. This indicates that we can approximate the matrix $\mathbf{A}$ using just two vectors (the first column of $\mathbf{U}$ and the first row of $\mathbf{V}^{\mathrm{T}}$) and the first entry on the diagonal of $\mathbf{S}$ and expect to produce a reasonable approximation because 88.6% of the variability in the data will be captured by the $r = 1$ approximation:[2]

$$\begin{pmatrix} 0.95 & 0.37 & 0.58 & 0.18 \\ 0.59 & 0.15 & 0.40 & 0.11 \\ 0.29 & 0.22 & 0.11 & 0.03 \\ 0.12 & 0.12 & 0.04 & 0.05 \\ 0.23 & 0.23 & 0.07 & 0.03 \\ 1.24 & 0.51 & 0.74 & 0.23 \end{pmatrix} \approx \begin{pmatrix} 0.554 \\ 0.341 \\ 0.169 \\ 0.076 \\ 0.139 \\ 0.723 \end{pmatrix} \big(2.141\big) \big(0.801 \ 0.328 \ 0.478 \ 0.148\big)$$
$$\mathbf{A} \qquad\qquad \mathbf{U}_1 \qquad\quad \mathbf{S}_1 \qquad\qquad \mathbf{V}_1^{\mathrm{T}}$$

$$= \begin{pmatrix} 0.95 & \mathbf{0.39} & \mathbf{0.57} & 0.18 \\ 0.58 & \mathbf{0.24} & \mathbf{0.35} & 0.11 \\ 0.29 & \mathbf{0.12} & \mathbf{0.17} & \mathbf{0.05} \\ \mathbf{0.13} & \mathbf{0.05} & \mathbf{0.08} & \mathbf{0.02} \\ \mathbf{0.24} & \mathbf{0.10} & \mathbf{0.14} & \mathbf{0.04} \\ 1.24 & 0.51 & 0.74 & 0.23 \end{pmatrix} .$$

---

[2] The SVD of the matrix is written only to three decimal places to conserve space. In practice we would use more digits.

In the above approximate version of $\mathbf{A}$ the entries that are not correct to 2 decimal places are written in bold. Examining this approximation we see that one row is exactly reproduced, and the first column has no errors larger than 0.01. Depending on our application, this approximation may be sufficient. In such a case we would only consider a single variable for each case as stored in the vector $\mathbf{U}_1$, for our data matrix, and then use $\mathbf{S}_1$ and $\mathbf{V}_1$ to reproduce the original variables in $\mathbf{A}$.

If we use the two term approximation, $r = 2$, to approximate $\mathbf{A}$, we get nearly a perfect approximation:

$$
\underbrace{\begin{pmatrix} 0.95 \ 0.37 \ 0.58 \ 0.18 \\ 0.59 \ 0.15 \ 0.40 \ 0.11 \\ 0.29 \ 0.22 \ 0.11 \ 0.03 \\ 0.12 \ 0.12 \ 0.04 \ 0.05 \\ 0.23 \ 0.23 \ 0.07 \ 0.03 \\ 1.24 \ 0.51 \ 0.74 \ 0.23 \end{pmatrix}}_{\mathbf{A}} \approx \underbrace{\begin{pmatrix} 0.554 \ -0.101 \\ 0.341 \ -0.443 \\ 0.169 \ \ 0.516 \\ 0.076 \ \ 0.323 \\ 0.139 \ \ 0.650 \\ 0.723 \ \ 0.007 \end{pmatrix}}_{\mathbf{U}_2} \begin{pmatrix} 2.141 \ \ \ 0 \\ 0 \ \ \ 0.233 \\ S_2 \end{pmatrix}
$$

$$
\times \underbrace{\begin{pmatrix} 0.801 \ \ 0.328 \ \ 0.478 \ \ \ \ 0.148 \\ -0.047 \ 0.865 \ -0.496 \ -0.061 \end{pmatrix}}_{\mathbf{V}_2^T} = \begin{pmatrix} 0.95 \ 0.37 \ 0.58 \ 0.18 \\ 0.59 \ 0.15 \ 0.40 \ 0.11 \\ \mathbf{0.28} \ 0.22 \ 0.11 \ \mathbf{0.05} \\ \mathbf{0.13} \ 0.12 \ 0.04 \ \mathbf{0.02} \\ 0.23 \ 0.23 \ 0.07 \ 0.03 \\ 1.24 \ 0.51 \ 0.74 \ 0.23 \end{pmatrix}.
$$

From this result we see that the largest error is 0.03, and nearly all of the entries are exact to 2 decimal places. Therefore, we can replace the 4 measurements in the original data with 2 linear combinations of the original columns, given by the first two columns of $\mathbf{V}$. To see how such a reduction could be used in practice, we examine a case study next.

Data Reduction with SVD
When we replace a matrix by its ℓ term SVD, we reduce the storage required. If the original matrix was of size $I \times J$, it required IJ numbers in storage/memory. For an ℓ term SVD we need to store $I\ell$ numbers for $\mathbf{U}_\ell$, ℓ numbers for $\mathbf{S}_\ell$ because only the non-zero terms need to be stored, and $J\ell$ numbers for $\mathbf{V}_\ell$. Therefore, the total storage for the ℓ term SVD is $\ell(I + J + 1)$. Therefore, if ℓ is much less than I or J, the storage is greatly reduced.

Consider the case of having $I = 10^6$ with $J \ll I$. This is a common case in data applications where there might be millions of cases of tens to hundreds of measurements. In this case the storage required in the ℓ term SVD is $\ell(I + J + 1) \approx \ell I$. The fraction of data needed to be stored, relative to the original matrix, is then $\ell I / IJ = \ell/J$. As a result, the storage is reduced by a factor ℓ/J. It is commonplace that ℓ could be an order magnitude smaller than J so that the data storage requirement goes down by about an order of magnitude.

Singular Value Decomposition

The singular value decomposition (SVD) is a technique to find structure and reduce the size of data by changing a set of correlated variables into a smaller set of uncorrelated variables. It involves placing the data into a matrix where each case is a row and the columns are the set of measurements related to that case. Upon performing the SVD we obtain the following insights into our data:

- The SVD reveals how many truly independent variables are there based on the number of non-zero singular values, r.
- Using only a few singular values we can represent the original data using a different set of uncorrelated measurements. For $\ell < r$ the accuracy of this reduced representation depends on how close to 1 the value of m_ℓ is.
- The amount of memory required to store the data can be reduced from $I \times J$ to $\ell(I + J + 1)$ by using an ℓ term SVD.

4.2   Case Study: SVD to Understand Time Series

We consider a time series that contains the temperature as a function of time for the simulation of an experiment. The experiment involves a series of laser beams focused on the inner walls of a cylindrical container made of gold, called a hohlraum. The hohlraum radiates x-rays that heat a center chamber. The goal of the simulations is to find a configuration that has a high temperature for a long period of time so that a sample can be heated to determine the properties of the sample at high temperature. The simulations have varied the size and dimensions of the container and the laser pulse length. Eventually, the goal is to use supervised learning to predict the temperature as a function of time, but it would be difficult to predict the temperature at each time. We will use the SVD to reduce the temperature as a function of time to a set of temperature profiles. Each time series of temperatures is then represented as a linear combination of these profiles. Supervised learning will then be used to predict the coefficients of this linear combination.

The hohlraum is shown in Fig. 4.1. In the figure the center chamber is where the sample (shown as a rectangle) is placed to be heated, and the dimensions of the nominal configuration are shown. In the simulations the laser is focused in the regions outside the baffles that separate the center chamber from the outer region.The geometry of the hohlraum is varied through the definition of three parameters. The parameter "scale" multiplies all measurements to adjust the overall size of the geometry; "scl" multiplies Z_{baf} and adjusts R_{apt} so that the relationship

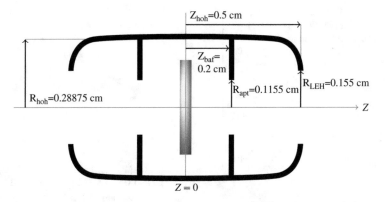

Fig. 4.1 Schematic of nominal hohlraum geometry for the simulated experiments. The sample to be irradiated is shown for illustration purposes in the center of the geometry

Fig. 4.2 The mean temperature of the gold as a function of time for the 15 simulations in the data set. This figure and subsequent figures make use of natural cubic splines to connect the data points

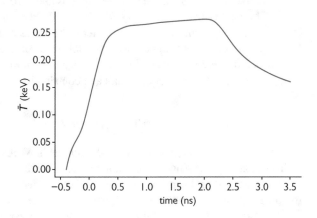

$R_{apt}/Z_{baf} = R_{hoh}/Z_{hoh}$ holds. The variable "rapt" multiplies R_{apt}. Each of these three parameters, scale, scl, and rapt, can take values of 0.8, 0.85, 1, 1.05, 1.2, 1.25.

We have data from 15 simulations for the average temperature of the gold near where the laser is focused at 30 different times from -0.399979 to 3.50014 ns; the zero time is when the main part of the laser pulse begins. We assemble this data into a matrix of 15 rows and 30 columns. We then compute the mean of each column of the matrix to get the average temperature profile over all of the simulations. This mean temperature response is shown in Fig. 4.2; the units of temperature used are keV where $1\,\mathrm{keV} \approx 11.6 \times 10^{6}$ K. Before computing the singular value of the data matrix the mean of each column is subtracted from the data matrix so that each column represents the difference from the mean for each simulation as a function of time. We call this mean temperature profile $\bar{\mathbf{T}}$.

We now compute the SVD of the data matrix, after subtracting the mean. The values for m_{ℓ} from the SVD indicate that the first few singular values capture nearly all the variations between the different simulations. The value for m_1 is nearly 0.7,

Fig. 4.3 The first two
singular vectors for the
simulation data (i.e., the first
two columns of **V**)

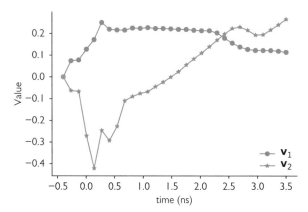

implying that the largest singular value explains about 70% of the variation between
cases. The next two values are $m_2 = 0.836$ and $m_3 = 0.882$. Given that we care
most about the overall behavior of the temperature as a function of time, the first
two singular components appear to be sufficient to capture the temperature profile.
The means that we can write the vectors of temperature for each case i at the 30
time points, **T**, as the sum of the mean and contributions from the first two columns
of **V**:

$$\mathbf{T}_i \approx \bar{\mathbf{T}} + u_{i1} S_{11} \mathbf{v}_1 + u_{i2} S_{22} \mathbf{v}_2, \tag{4.5}$$

where $\mathbf{v}_1$ and $\mathbf{v}_2$ are the first and second columns of **V**, respectively.

To visualize these singular components we plot the first two columns of **V** to see
what they represent. From Fig. 4.3 we see that the first singular vector corresponds
to how far above or below a simulation case is to the mean response: a case with a
positive value in the first column of **U**, that is, u_{i1} is positive, will have its overall
profile above the mean response, $\bar{\mathbf{T}}$. Conversely, if u_{i1} is negative, case i will have
temperatures that are below the mean values. The second column of **V**, $\mathbf{v}_2$ is a bit
harder to interpret because it has positive and negative components. By using the
fact that we are considering adjustments to the mean temperature, we can see that
a case with a large value for u_{i2} will have temperatures that tend to be lower early
in time and higher later in time, relative to the mean and the contribution from the
first singular vector. This also means that if the value u_{i2} is negative, the early time
temperatures are higher and the later time temperatures are lower: such a scenario
corresponds to a simulation that heats up faster and cools down faster.

To further understand the singular value decomposition we plot the cases with
the largest positive and negative values for u_{i1} in Fig. 4.4a. Comparing these cases
to the mean response we see that shifting the results above or below the mean using
the value of u_{i1} can capture the differences between these cases and the mean. When
we look at these cases we see that case 4 is the largest hohlraum, and case 1 is the
smallest hohlraum that we have simulation data for. Having the largest geometry
that gives temperatures below the mean makes physical sense because when the

Fig. 4.4 The approximation of different cases using the first two singular values. Cases 4 and 1 had the largest positive and negative magnitudes, respectively, for u_{i1}; cases 5 and 9 had the largest positive and negative magnitudes, respectively, for u_{i2}. The full data is shown using a solid black line and the approximations using 1 and 2 singular values use a dashed and dotted line, respectively. (**a**) Cases 1 (upper curve) and 4 (lower curve). (**b**) Cases 5 and 9

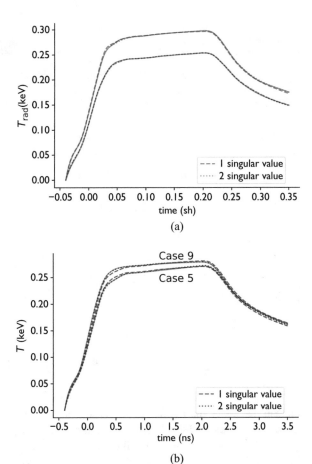

geometry is larger there is a larger volume to contain the energy and the energy density goes down. A smaller geometry gives a higher energy density and a higher temperature.

The impact of the contribution from u_{i2} is shown in Fig. 4.4b. Here we show the two cases with the largest positive and negative values for u_{i2}: cases 5 and 9, respectively. For these cases an approximation based solely on the first singular value cannot capture the behavior at early and late times near the plateau of the temperature profile. When the second singular values are included, the approximate profiles are indistinguishable from the true profiles. We can also see that the influence of the second singular component is much smaller than the first singular component, as expected from the difference between m_1 and m_2. Cases 5 and 9 had the extreme values for the sample chamber length: case 5 had the minimum value for this quantity in the data, and case 9 had the maximum value. From the figure it appears that having a large value of sample chamber length makes the temperature

Fig. 4.5 Illustration of the linear models used to predict the values of u_{i1} and u_{i2} as a function of the geometric parameters

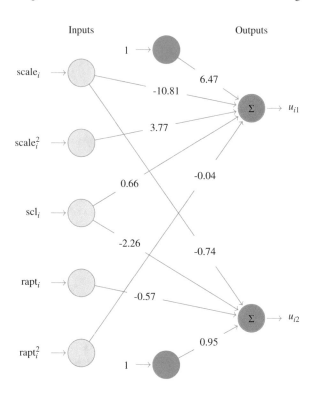

plateau rise faster and fall off sooner; case 5 has a small value for this parameter and the plateau is reached at a later time and remains there longer.

Using the results above we will build two linear regression models, as described in Sect. 2.2, to predict the values of u_{i1} and u_{i2} as a function of the three parameters via Eq. (4.5). This will allow us to predict the time profile of the temperature for any given modification to the hohlraum that can be described via the parameters scale, scl, and rapt. After some investigation (including cross-validation) we find that adequate models are given by

$$u_{i1} = 6.47 - 10.81(\text{scale})_i + 3.77(\text{scale})_i^2 + 0.66(\text{scl})_i - 0.04(\text{rapt})_i^2, \qquad (4.6)$$

$$u_{i2} = 0.95 - 0.74(\text{scale})_i - 2.26(\text{scl})_i - 0.57(\text{rapt})_i. \qquad (4.7)$$

The linear models are visualized in Fig. 4.5. From these models we can see that for u_{i1} the most important parameter is the scale parameter. This is consistent with our findings from the above where the simulation cases with extreme values of scale had extreme values for u_{i1}. For u_{i2} it appears that scl is the parameter that has the largest impact.

We can compare the predicted temperature profiles with the data used to train the model. When doing so we find that on a graph the model output and the data are

Fig. 4.6 Prediction of the crashed simulation with scale $= 0.8$, scl $= 1$, and rapt $= 1$

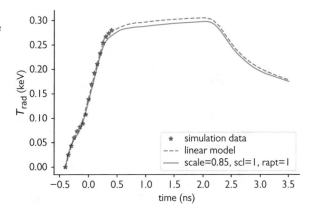

indistinguishable. It is more interesting to do a blind prediction on data the model did not use to train it.

To test the model we can compare it with a simulation that did not complete. The simulation with scale $= 0.8$, scl $= 1$, and rapt $= 1$ crashed before it could complete. The data the simulation produced was not used in the construction of the SVD or the linear models. Using the linear models we can predict how this simulation would have proceeded. The simulation data that is available and the prediction from the linear models for u_{i1} and u_{i2} used in Eq. (4.5) are shown in Fig. 4.6. From the figure we see that the prediction tracks the known data. We can also see in the figure that this predicted temperature profile is higher than the simulation that had scale $= 0.85$, scl $= 1$, and rapt $= 1$, as expected from the model. This is just one prediction that we could make using our linear model. In principle we could predict the temperature profile for any geometry that can be described using the three parameters, though we would want to be careful for values of these parameters outside the range used in our training data.

Application of the Singular Value Decomposition

When we apply the SVD to data, we can gain insight into the data by visualizing the singular component vectors and by looking at cases with extreme values of u_{ij}. After reducing the data with the SVD, we can use the values of u_{ij} as dependent variables in supervised learning models. This can reduce the burden of the number of models to fit by reducing the number of dependent variables.

4.3 K-means

In many analysis tasks, we are interested in natural groupings, or clusters, in the data. For example, if we looked at all the different models of cars made in a given year, and for our variables we looked at the number of doors, the engine displacement, the vehicle weight, and other similar characteristics, we would find that there are different groupings in the data that naturally arise. These clusters could include sports cars with large engines per weight, compact cars with high fuel efficiency, and light trucks with large vehicle weight and engine displacement and typically a smaller number of doors. By looking for clusters in data we are looking to find similar groupings to this automotive example: are there ways to understand a given case in the data set as part of particular group.

One example of natural groupings in data can be seen in baseball statistics. Fans of the sport will know that there are different types of offensive players. These types include power hitters that hit a large number of home runs and doubles, fast runners who hit singles and steal bases, with the occasional triple, and finally all around hitters that excel at power hitting and get on base frequently. If one looks at data from baseball statistics, these natural groupings (and others) can be found in the data.

One way to determine natural groupings, or clusters, in the data is through the K-means algorithm. To define this algorithm we consider a data matrix $\mathbf{X}$ comprised of I observations of J variables. We write each row of $\mathbf{X}$ as $\mathbf{x}_i = (x_{i1}, \ldots, x_{ij}, \ldots, x_{iJ})$. The goal of K-means is to assign each observation to one of the K classes denoted as S_k for $k = 1, \ldots, K$ where the following loss function is minimized:

$$L = \sum_{k=1}^{K} \sum_{\mathbf{x}_i \in S_k} \|\mathbf{x}_i - \boldsymbol{\mu}_k\|^2, \tag{4.8}$$

where $\boldsymbol{\mu}_k$ is the vector of the mean each of the variables of the observations in set S_k:

$$\boldsymbol{\mu}_k = \frac{1}{N_k} \sum_{\mathbf{x}_i \in S_k} \mathbf{x}_i. \tag{4.9}$$

When the minimum of L is found, we have each case assigned to a group, S_k, and each group is represented by the mean value of the group. Therefore, we can think of each group, or cluster, as observations "nearby" the mean of the group.

This is a useful way to understand the different clusters that arise in the data, but there are reasons that these clusters may not be ideal. For instance, just because we have minimized L in Eq. (4.8) does not mean that all the cases are close to a group mean. Moreover, in minimizing L we have to pick the number of clusters K. The value of the loss function will likely change with K, and it may be difficult to know ahead of time what K should be.

Another characteristic of K-means clustering is that the algorithms used in practice to minimize the loss function in Eq. (4.8) do not provide the global minimum and can converge to a local minimum. This means that performing the K-means clustering repeatedly may give different results. For these reasons one should check the run-to-run variability in the results of the K-means clustering to make sure there are not issues with local minima.

There are also variations on the K-means algorithm. One simple variation is using the median of each cluster in the loss function rather than the mean. The resulting method is called the k-medians algorithm. There are also hierarchical versions that produce clusters that are subsequently split into smaller clusters so that grouping at different scales can be investigated.

K-means Clustering
K-means finds K different clusters in a set of data by grouping each case based on how close it is to the average of the cases in a cluster. This method requires the user to specify the number of clusters to find, K. Also, the clusters found may change when the algorithm is run multiple times.

4.3.1   K-means Example

To demonstrate K-means clustering we consider data where two variables per observation are generated by sampling from a normal distribution with mean $\bar{x}_{ik}$ for $k = 1, 2,$ or 3 and standard deviation 0.6. For the initial data we have $\bar{x}_1 = (1, 0)$, $\bar{x}_2 = (1.75, 2)$, $\bar{x}_3 = (4, 0.5)$. There are 51, 49, and 52 samples from the three distributions, respectively. Using these random samples, we cluster the data using K-means with $K = 3$. Our data has three natural clusters from the three distributions; we would like K-means to find these three clusters.

The original data and the results of the K-means clustering are shown in Fig. 4.7. In the top panel the original samples are shown and the distribution a point was sampled from is indicated by a different marker and color. We can see in the data that there is no overlap in the samples, that is, there appears to be a separation between the clusters. In the lower panel of Fig. 4.7 the cluster that each point belongs to is identified by the marker and color. Also, the cluster centers identified by K-means are drawn as a large point. Those points that K-means associates with a different cluster than the distribution used to generate the point are shown with no fill as open symbols. Here we see that there are two points in the original data set that are closer to a center identified by K-means than the center of the distribution it was drawn from. This is not a flaw of K-means; it just means that these two points were sufficiently far from the center of their distribution, due to random chance, that they look more like belong to a different cluster.

Fig. 4.7 K-means clustering example with separated clusters. (**a**) Original data with different markers for each cluster, and (**b**) the clusters identified by K-means with $K = 3$. The center of each cluster chosen by K-means is indicated with a large marker, and clustered points identified by K-means that disagree with the distribution the original data was drawn from are plotted using an unfilled symbol. (**a**) Original data. (**b**) Clusters from K-means

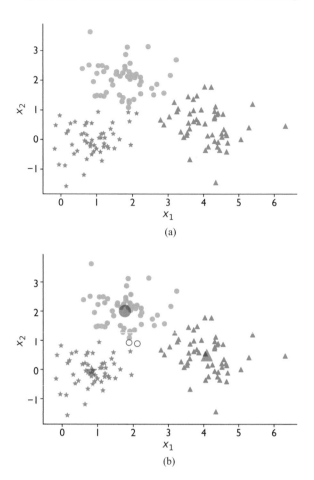

The centers of the clusters as identified by K-means are $\bar{\mathbf{x}}_1 = (0.825, -0.0525)$, $\bar{\mathbf{x}}_2 = (1.761, 2.003)$, $\bar{\mathbf{x}}_3 = (4.083, 0.491)$. These values are near the true centers of the distributions used to produce the data. It is worthwhile to remember that the K-means algorithm has no information about the original distribution, and we only told it to look for 3 clusters. Given how it has performed, we could have reasonable confidence that a new point generated from one of the distributions could be correctly identified as which distribution it came from.

We can make the clustering problem more difficult by increasing the standard deviations of the distributions so that there is more overlap in the data. Specifically, we increase the standard deviation in the normal distributions to be 0.75. This causes the samples from each distribution to significantly overlap, as shown in the top panel of Fig. 4.8. We would not expect K-means to give identical answers as before because there are many points that are closer to the center of a different distribution than the center of the distribution it was sampled from. However, the centers of the

Fig. 4.8 K-means clustering example with overlapping clusters. (**a**) Original data with different markers for each cluster, and (**b**) the clusters identified by K-means with $K = 3$. The center of each cluster chosen by K-means is indicated with a large marker, and points clustered by K-means that disagree with the original data are drawn as an outline

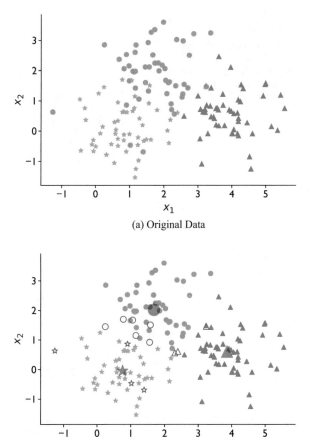

(a) Original Data

(b) Clusters from K-means

distributions are not changed so it may be possible to identify these clusters using a centroid.

In the lower panel of Fig. 4.8 we see that there are many more points identified with a different cluster than the distribution it was generated from. This comes from the fact that K-means only uses distances in its clustering. The cluster centers identified by K-means are $\bar{x}_1 = (0.752, -0.046)$, $\bar{x}_2 = (1.679, 2.018)$, $\bar{x}_3 = (3.855, 0.605)$. These centers are less accurate than what we saw when the clusters were better separated, but they still are reasonable approximations and qualitatively they indicate the three clusters where we would expect them to be.

In a sense we cheated a little bit in this example. The primary user input to K-means is the value of K. We used $K = 3$ so far in the example, and K-means found our three clusters without much difficulty. The question naturally arises of how would K-means behave if the number of clusters was run with a different value of K. The results using $K = 2$ and $K = 5$ on the data using a standard deviation

Fig. 4.9 K-means clustering example with separated clusters using different values of K. The center of each cluster chosen by K-means is indicated with a large marker. (**a**) $K = 2$. (**b**) $K = 5$

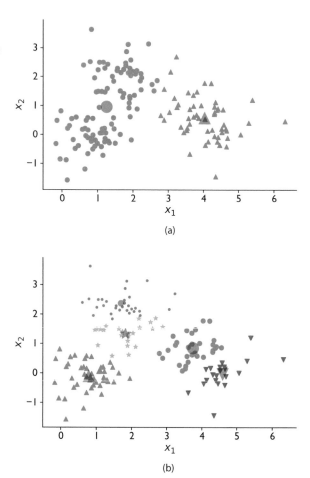

of 0.6 in the distributions used to generate the samples are shown in the top and bottom panels of Fig. 4.9, respectively. In the figure we see that when $K = 2$ the two clusters identified seem to be divided solely on the value of x_1. This is likely due to the fact that the range of x_1 in the data is greater than the range in x_2. Just viewing the data in the figure this clustering seems reasonable, but it is possible to spot the problem. In the left cluster the center is not near many data points: it appears that there are two clusters that this point is splitting. This is one way to identify that there may be a problem. We can also look at the value of the loss function, Eq. (4.8). The value of L for $K = 2$ is 226.5 and for $K = 3$ it is $L = 101.9$. The magnitude of the loss function is smaller for $K = 3$, but we should look at the value of L per cluster by multiplying by K. This will help deal with the fact that if $K = I$, that is, the number of clusters is equal to the number of data points, the value of $L = 0$. For this example $LK = 453.0$ for $K = 2$ and $LK = 305.6$ for $K = 3$. Based on this measure we can see that $K = 3$ is superior to $K = 2$.

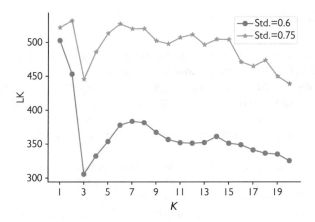

Fig. 4.10 The value of the loss function, L, times the value of K used in the K-means clustering for the two data sets considered

The case of $K = 5$ as shown in the lower panel of Fig. 4.9 has the resulting clusters appearing to split two of the three "true" clusters into two different clusters. As with the $K = 2$ case, the clusters identified seem reasonable to the eye at first look. The value of LK for $K = 5$ is 353.6, significantly higher than the value of $K = 3$. This would give us a reason to select $K = 3$ over $K = 5$.

We can further investigate the idea of using LK to choose K by examining these values for both the data sets we have generated (one using 0.6 as the standard deviation in the normal distributions and the other using 0.75 for the standard deviation). The value of LK is plotted as a function of K in Fig. 4.10 for each of these data sets. In both curves there is an obvious local minimum at $K = 3$. Though it appears that large values of K will have a lower value of LK, if we use parsimony as a guide, we can easily settle on $K = 3$ rather than some value $K > 20$. Looking at a plot such as this one helps guide the selection of K.

Tips for Applying K-means Clustering

When using K-means, one should try several different values of K and look at the loss times the number of clusters: a low value for this product indicates a better clustering. Using this metric, and the principle of parsimony, will help in the selection of K that is large enough to describe the clustering of the data, without adding unnecessary clusters.

For data that have a small number of dimensions, that is J is small, one can visualize the clusters. One can inspect if the cluster centers are close to many data points. If the cluster centers are in a location with a small density of points, more clusters may be needed.

4.4 t-SNE

When we consider finding structure in data, one difficulty is that in high-dimensional
data we cannot easily visualize how the structure arises. It would be beneficial
to have a way to spot clusters in a high-dimensional data set and then use that
knowledge to inform our approach. Visualizing data in 3-D is hard enough, and
the situation with tens or hundreds of independent variables seems impossible.
To address this issue t-distributed stochastic neighbor embedding (t-SNE) was
developed [1].

The idea behind t-SNE is to look for groupings in the high-dimensional data set
and then map it into a 2-D or 3-D distribution for visualization. The "t" in t-SNE
does a lot of the work in making the structure of the data stand out when we look
at the data in low dimensions. The t-distribution with a single degree of freedom,
also known as the Cauchy distribution, has long tails, i.e., the distribution goes to
zero very slowly away from the mean. To see this we will compare the probability
density function for a normal distribution (or Gaussian distribution) with a Cauchy
distribution. The probability density function for a normal distribution for a single
random variable x is given by

$$f(x|\mu, \sigma) = \frac{1}{\sqrt{2\pi\sigma^2}} e^{-\frac{x-\mu)^2}{2\sigma^2}}, \tag{4.10}$$

where $f(x|\mu, \sigma)$ is the probability density for x given a distribution mean of μ and
standard deviation σ. The Cauchy distribution (i.e., the t-distribution with a single
degree of freedom) has a probability density function given by

$$f(x|\mu, \gamma) = \frac{1}{\pi\gamma} \left(\frac{\gamma^2}{(x-\mu)^2 + \gamma^2} \right). \tag{4.11}$$

For the Cauchy distribution μ is also the mean of the distribution. The parameter γ
gives information about the tails of the distribution, but the Cauchy distribution has
an undefined standard deviation and variance because the distribution approaches
zero so slowly outside of the mean. In Fig. 4.11 normal and Cauchy distributions
are compared using $\mu = 0$ and $\sigma = \gamma = 1$. The tails of the Cauchy distribution
are much heavier than that for the normal as shown by the distribution approaching
zero more slowly as x goes away from the mean. If we look at the value at $x = 4$,
we get that the value for the Cauchy distribution is about 140 times greater than that
for the normal.

t-SNE uses both of these distributions: a normal distribution is used to find the
clusters in the high-dimensional space, and these clusters are then mapped into a
Cauchy distribution in 2-D or 3-D to visualize the data. The normal distribution is
used in the high-dimensional space because it is easy to work with and has nice
properties when used on large data sets. The Cauchy distribution is used for the

Fig. 4.11 Comparison of
normal (blue) and Cauchy
distributions (red) with the
same mean, $\mu = 0$, and
$\sigma = \gamma = 1$. Notice that the
Cauchy distribution
approaches zero much more
slowly as x gets farther from
the mean

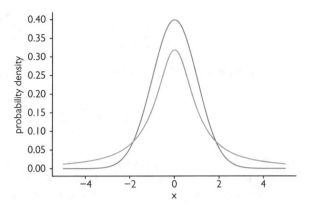

visualization because the long tails tend to spread out the clusters so that they are
easier to visualize.

> **t-SNE**
> High-dimensional data sets, such as images, are difficult to cluster and
> visualize the clusters. t-SNE maps clusters found in a high-dimensional data
> set to 2-D or 3-D in such a way that clusters can be visualized. The trick
> it uses is to find groupings in the high-dimensional space using a normal
> distribution and then map that distribution into t-distribution with one degree
> of freedom (also called the Cauchy distribution) in 2-D or 3-D. Using the
> Cauchy distribution in the low-dimensional space helps to spread out the
> clusters so that they can be seen visually.

4.4.1 Computing t-SNE

Note: This section involves the manipulation of multidimensional probability
density functions and may be skipped for readers without previous experience in
these functions.

The first step in t-SNE is to determine the clusters in the high-dimensional space.
We will denote each data point as $\mathbf{x}_i$ for $i = 1, \ldots, I$. The method defines the
similarity between two points $\mathbf{x}_i$ and $\mathbf{x}_j$ as the conditional probability that point
$\mathbf{x}_i$ would select $\mathbf{x}_{i'}$ as its neighbor if the probability of selecting a neighbor was
a normal distribution with mean $\mathbf{x}_i$ and standard deviation σ_i. This conditional
probability density is written as $p_{i'|i}$ and is the probability that point $\mathbf{x}_{i'}$ would be
selected as a neighbor given that $\mathbf{x}_i$ were selecting its neighbor:

$$p_{i'|i} = \frac{\exp\left[-\|\mathbf{x}_i - \mathbf{x}_{i'}\|^2/(2\sigma_i^2)\right]}{\sum_{k \neq i} \exp\left[-\|\mathbf{x}_i - \mathbf{x}_{i'}\|^2/(2\sigma_i^2)\right]}, \tag{4.12}$$

where we are using the Euclidean norm to measure distance,

$$\|\mathbf{x}_i - \mathbf{x}_{i'}\|^2 = \sum_{j=1}^{J} (x_{ij} - x_{i'j})^2. \tag{4.13}$$

This conditional probability has a form that makes it so that the further the distance between two points, the smaller the probability that they are neighbors. The distance is measured in units of σ_i. The value of σ_i is chosen based on the data so that σ_i is larger in regions where there are many points and smaller in regions where there are few data points. In implementations of t-SNE one sets a parameter called the perplexity to control how this density of points is determined.

Once we have computed the conditional probabilities $p_{i'|i}$ we want to create an equivalent low-dimensional distribution. We use $q_{ii'}$ to denote the probability density in this low-dimensional distribution and $\mathbf{y}_i$ to denote the points in the low-dimensional space where $\mathbf{y}_i = (y_{i1}, y_{i2})$ or $\mathbf{y}_i = (y_{i1}, y_{i2}, y_{i3})$ depending on if the low-dimensional space is 2- or 3-dimensional. We call the number of dimensions d. The probability of $\mathbf{y}_i$ and $\mathbf{y}_{i'}$ being neighbors is based on the multidimensional Cauchy distribution

$$q_{ii'} = \frac{\left[1 + \|y_{ik} - y_{i'k}\|^2\right]^{-1}}{\sum_{\ell=1}^{I} \sum_{\ell'=1}^{I} \left[1 + \|y_{\ell k} - y_{\ell'k}\|^2\right]^{-1}}. \tag{4.14}$$

Using the Cauchy distribution for $q_{ii'}$ allows the numerator to go to zero more slowly than using an exponential (as the normal distribution does). This means that the difference between points far away is more likely to have a large value in the low-dimensional space, and therefore the distribution can tell the difference between points once they are far away. Or, to think about it another way, the long tails of the Cauchy distribution allow it to push points farther away in the low-dimensional space than they would be in the high-dimensional space. This helps to separate the clusters.

The question arises how do we pick the points $\mathbf{y}_i$. To find these values we ask that the distributions in the high-dimensional space, $p_{i'|i}$ and $q_{ii'}$, match in some sense. To make these distributions match we minimize the loss function

$$L = \sum_{i=1}^{I} \sum_{i'=1}^{I} p_{ii'} \log \frac{p_{ii'}}{q_{ii'}}, \tag{4.15}$$

where $p_{ii'} = 0.5(p_{i'|i} + p_{i|i'})$. This is called the Kullback–Leibler (KL) divergence. Due to the logarithm in the formula, the minimum value of L occurs when all $q_{ii'} =$

Table 4.1 The 10 classes of the Fashion MNIST data set

Class	0	1	2	3	4	5	6	7	8	9
Description	T-shirt/top	Trouser	Pullover	Dress	Coat	Sandal	Shirt	Sneaker	Bag	Ankle boot

$p_{ii'}$. The values of $\mathbf{y}_i$ are chosen so that L is minimized. This then gives us the values for $\mathbf{y}_i$ that are distributed by a Cauchy distribution that matches the high-dimensional normal distribution. The minimization of L is accomplished by using standard optimization techniques. These optimization techniques often rely on a parameter called the learning rate that we will discuss in some detail later.

4.4.2   Example of t-SNE

To demonstrate how t-SNE works we consider an image recognition data set called the Fashion MNIST image data set [2]. This is a set of 28×28 grayscale images that contain one of ten classes of clothing and fashion accessories. The classes are given in Table 4.1. Figure 4.12 contains example images from this data set for each of the 10 classes.

Though in science and engineering we rarely deal with fashion items, this set of images is related to the important problem of image classification. Whether we are considering an autonomous vehicle, drone flight, or other automated tasks, object recognition is an important problem. In our current study we ask if there are natural groupings in the data that t-SNE can uncover. We treat each image as having $28^2 = 784$ degrees of freedom, i.e., $J = 784$, where each value is the grayscale value of a pixel. In this manner we convert a set of images into a standard data matrix of size $I \times J$ where I is the number of images. In the application of t-SNE the algorithm has no information about the class of images, just the raw pixel data.

We apply t-SNE using the scikit-learn implementation on a sample of 6000 images from this data set. Before performing t-SNE we use an SVD on the data set with $\ell = 200$ to reduce the 784 degrees of freedom in the original data. This makes the t-SNE algorithm run faster because the step of computing the normal distribution will be over a smaller dimensional data set. We then apply t-SNE with $d = 2$ to produce a 2-D clustering; the additional parameters we set are a perplexity of 40 and a learning rate of 1200 (see Sect. 4.4.1).

The resulting clustering by t-SNE is shown in Fig. 4.13 where the location of the points is determined by t-SNE but we label the points by the class the point belongs to. The resulting figure looks like the map of an archipelago with several large islands with several small points in between. Using the map analogy we refer to compass directions with north pointing to the top of the figure. At the top left (north west) of the figure is an island of pants, and we see that there are few pant points that are not on this island. From this we can see that trousers/pants are easy for t-SNE to distinguish from the other images. We also notice that at the bottom of the figure there is a region of bags.

Fig. 4.12 Examples of images from each class of the Fashion MNIST data set

There are several other large islands that we can identify that contain several classes. There is a large island in the lower part of the figure that contains two larger regions connected by a thin strip. This island contains primarily shirts, pullover shirts, and coats in its north-west part and contains shirts, pullover shirts, and T-shirts/tops in the south-east part. On the right side of the figure we see an island of sneakers and sandals: to t-SNE these images are similar. In the center of the figure there is an island containing nearly all of the dresses in the western part, and several other classes in the eastern part.

·	T-shirt/top
●	Trouser/pants
▲	Pullover shirt
+	Dress
▼	Coat
★	Sandal
►	Shirt
◄	Sneaker
×	Bag
⋎	Ankle boot

Fig. 4.13 Resulting cluster from t-SNE on the Fashion MNIST data set

To get further insight into the clustering we can put the images at the locations identified by t-SNE; this is shown in Fig. 4.14. In this figure we do not plot every data point to avoid overlapping. Plotting the actual images can give us some insight into how the islands were formed. For instance we can directly see why the island with the two large ends connected by a thin strip was formed. This island contains images of roughly the same shape but the two ends of the island separate darker items from lighter images. We can see this effect in other islands as well (e.g., the center island). We also notice that t-SNE found two different types of boots. It looks like one has more prominent heels than the other.

By applying t-SNE to this data we can understand how the images themselves have different structures beyond the classifications that we have imposed. For instance, based solely on the data a light-colored sandal is more similar to a light-colored low-top sneaker than that sneaker is to a dark high-top sneaker. These insights can be used in supervised learning problems to understand how to pose supervised learning problems and to gain better understanding of high-dimensional data. For instance, if we could normalize the data so that only the shape of the object was important and the differences in brightness/lightness were removed, it might be easier to differentiate items. Or, it could indicate that the classification that we give to data points may not make the most sense as a function of the raw data.

Fig. 4.14 Rendering of the t-SNE clustering where the points are replaced by the image. Not all images are shown to avoid overlapping

> **Interpreting t-SNE**
> The results from a t-SNE clustering can tell us information about the data that
> may not be apparent beforehand. For instance, in the Fashion MNIST data if
> we classify images based on pixels alone it appears that light-colored shoes of
> different types are more similar than light shoes and dark shoes of the same
> type. This type of insight could be useful when devising supervised learning
> models that work with the data.

4.5   Case Study: The Reflectance and Transmittance Spectra of Different Foliages

There are many applications of using optical properties of an object to learn
about the nature of the object. One application is hyper-spectral imaging where
measurements are made using light over a wide range of wavelengths, including
outside of the visible range of 400–700 nanometers (nm). Hyper-spectral imaging
has lots of uses including detecting when crops are ready to be harvested. In this
case study we will examine the spectral properties of different leaves and connect
this to other information about the leaves such as chlorophyll content.

The data we use comes from the ANGERS leaf database [3]. This data set
includes optical and biochemical measurements of 276 different leaves spread
across 39 different species. For each leaf the reflectance and transmittance of the leaf
were measured for light at 2051 wavelengths from 400 to 2450 nm in increments
of 1 nm. The leaves measured were not evenly distributed across species, as shown
in Fig. 4.15, with the sycamore maple having 181 leaves in the data set. We are
interested how the different leaves in the data can be clustered based on their optical
properties.

In Fig. 4.16 the reflectance and transmittance as a function of wavelength are
shown for several leaves. The reflectance is the fraction of light at a particular
wavelength that is reflected, and the transmittance is the fraction that passes through
the leaf. The sum of the absorptance, the fraction of light that is absorbed at a
wavelength, plus the transmittance and reflectance equals 1. The underlying physical
properties of the leaf, including the amount of water in the leaf, the chlorophyll, and
other chemicals present, affect the reflectance and transmittance spectra. There is
also leaf to leaf variability in these spectra, as one would expect because the leaves
of a plant do not all look identical even to our eye.

We will now use our tools to investigate the natural structure in the reflectance
and transmittance data. We construct a data matrix where each row contains the
reflectance and transmittance spectra for the 2051 wavelengths measured for a total
size of 276×5102. We use the singular value decomposition to find a reduced
representation of this data and to remove any noise in the measurements, particularly

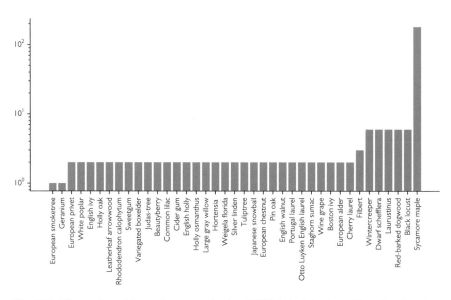

Fig. 4.15 The number of leaves by species in the ANGERS leaf database. Note that there is a large number of sycamore maple leaves relative to the other leaves

the long wavelength measurements above about 2500 nm. From the SVD we can look at the values of m_ℓ and determine what rank, r, we need to capture the leaf to leaf variability in the data set. The plot of m_ℓ is shown in Fig. 4.17. From this figure we can see that using just a few terms of the SVD we can capture a majority of the variability in the data. For instance, with $\ell = 5$ we can capture 85% of the variability in the data. Additionally, 90% of the variability can be captured with $\ell = 14$.

We choose to use a rank 4, i.e., $\ell = 4$, SVD approximation of the data matrix to find clusters in the data using K-means. When we look at the values of LK for this data, we find that 5 clusters is a reasonable number to search for. To visualize the 5 clusters we create scatter plots of the values of the columns in the matrix **U**, u_{ij}, for the first four columns $j = 1, \ldots, 4$ in Fig. 4.18. In this figure we plot all the pairs $(u_{ij}, u_{ij'})$ as the x- and y-axis, respectively. The marker and color for each point indicate what cluster it belongs. From this figure we notice that cluster 3 contains points that have a large value of u_{i3}, and that these points form a cluster that becomes obvious when visualizing the data in this form. Additionally, cluster 1 looks like it contains points that have large values of $u_{i1}, u_{i2},$ and u_{i4}.

Another way to understand the clusters that K-means reveals is to look at the reflectance and transmittance spectra that correspond to the center of each cluster, and the data point that is closest to that center. The center of each cluster is the four values of $\bar{u}_j$ for $j = 1, \ldots, 4$; to reconstruct the spectra we use Eq. (4.2) with $\ell = 4$ and where $\mathbf{U}_\ell$ is replaced with the row vector $(u_{i1}, u_{i2}, u_{i3}, u_{i4})$. The reflectance and the transmittance spectra for the centers of the clusters and the nearest leaf in the data are shown in Fig. 4.19.

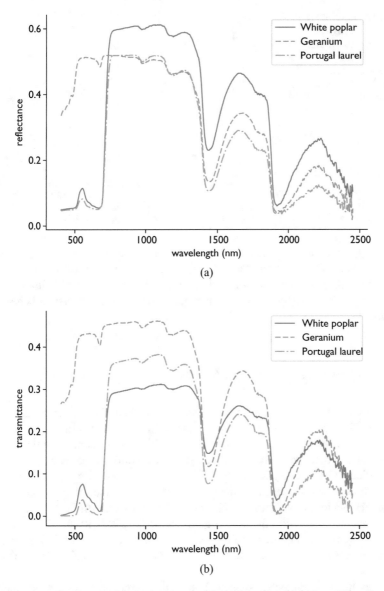

Fig. 4.16 Example reflectances and transmittances for leaf samples from three different species of plant. (**a**) Reflectance. (**b**) Transmittance

The spectra shown in Fig. 4.19 give further insight to the observations we made from Fig. 4.18. We notice that cluster 3 has a significantly elevated value for the reflectance and transmittance in the visible part of the spectrum on the left. This tells us that a large value for u_{i3} indicates that leaf i has high values in the visible range for reflectance and transmittance. In Fig. 4.19 we also observe that cluster 4

Fig. 4.17 Value of m_ℓ, c.f. Eq. (4.3), for the leaf reflectance and transmittance data. The symbols indicate the value of ℓ where $m_\ell = 0.85, 0.875, 0.9, 0.95,$ respectively

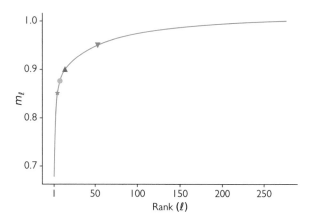

has a large value for the reflectance in the range of wavelengths above the visible, but a low value for the transmittance in this range. Leaves in clusters 2 and 5 appear to have the opposite behavior with a low reflectance and high transmittance in the wavelengths above visible; the difference between these clusters is their behavior in the visible range.

4.5.1   Reducing the Spectra to Colors

Though we have optical data for the leaves well beyond the visible range, we can compute the color of the reflected and transmitted light from a leaf as a visual representation of the spectra. To do this we will need a bit more optical theory. On a computer we typically represent colors as a combination of red, green, and blue called RGB. A color is a vector of three numbers that give the contribution of each color mixed together to get the final color; typically the values are normalized so that the vector has a maximum entry of 1. Each wavelength in the visible range can be mapped to a color; this was codified in the CIE 1931 RGB color space[3] [4]. This will allow us to convert the light reflected or transmitted at each wavelength into a color, but we still must determine how the colors mix to get the ultimate color that we would see.

To do this we assume that we shine light with a particular spectrum onto the leaf. If the light has a known spectrum, we can integrate the product of the light intensity at a particular wavelength, the color values at that wavelength, and the reflectance or transmittance value at that wavelength to get a vector corresponding to the RGB values for the reflectance or transmittance. For the light intensity we use a blackbody

---

[3]The International Commission on Illumination (CIE) used a series of experiments where observers matched light at a particular wavelength to colors that have a defined RGB value. The commission still exists, but it does not appear that its members are called the *Illuminati*.

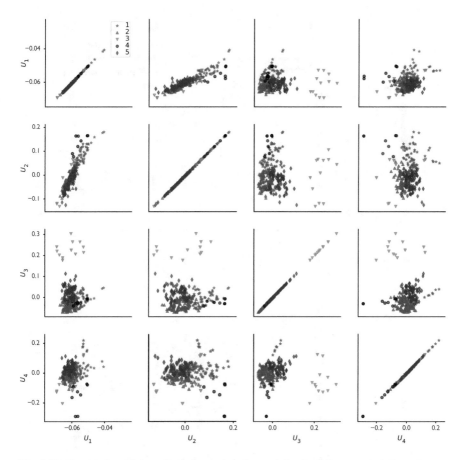

Fig. 4.18 Scatter plots of the values of u_{ij}, $j = 1, \ldots, 4$ for the reflectance and transmittance spectra for the leaf data set. The shape/color of the points indicates which of the 5 clusters the leaf belongs to

spectrum at a temperature of 5800 K; this is approximately the spectrum of sunlight at the earth's surface and is particularly relevant for leaf illumination.

The reflected and transmitted light colors for 40 different leaves are shown in Fig. 4.20. The figure further illuminates, so to speak, the observations we made from the hyper-spectral data. Cluster 3 has leaves that have a similar color for transmitted and reflected light and the color is more red/orange than the other leaves. It appears that these leaves were not particularly green when they were collected. This could indicate that the leaf or the entire plant was not healthy. It would be possible to use this result in remote sensing to identify distressed plants in the field: if satellite or drone surveys can find regions of a field where leaves fall into cluster 3, the plants in that region should be investigated for potential problems.

From the colors shown in Fig. 4.20 we can also see that cluster 5 contains leaves with more yellow/orange in the reflected and transmitted light than we observe in

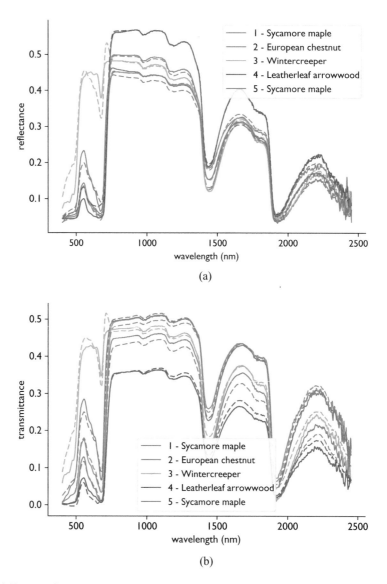

Fig. 4.19 The reflectance and transmittance spectra for the centers of the 5 clusters from K-means and the leaf nearest the center of the cluster. The dashed lines are the reconstructed spectra for the cluster centers, and the solid lines are the leaf closest to the cluster center. The plant name for the leaves closest to the center is also provided. (**a**) Reflectance. (**b**) Transmittance

clusters 1, 2, and 4. When we look at the spectral data in Fig. 4.19 we see that cluster 5 has higher values in the visible range than clusters 1, 2, and 4. Also, we notice that there are no sycamore maple leaves in clusters 3 and 4. This is interesting because there were a large number of leaves from this plant in the data. This indicates that it

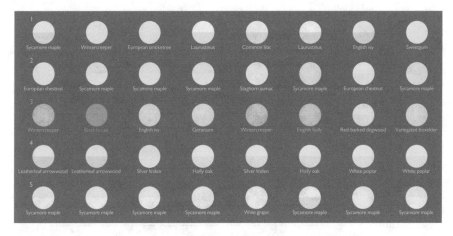

Fig. 4.20 Effective color of the reflected and transmitted light for the different clusters of leaves. Each row corresponds to a cluster, with cluster 1 being the top row. The columns contain the 8 leaves closest to the center of the cluster. Each circle has the top half color being the reflected light, and the bottom half being the transmitted light. The names of the plant are shown as well

may be possible to have high confidence that a leaf did not come from this plant if its spectra falls into cluster 3 or 4.

Beyond these observations it becomes somewhat difficult to describe the differences between the clusters using color information alone because we are reduced to using terms like greenish-yellow or pale green to describe colors. This is one of the benefits to studying the actual spectra rather than colors. Moreover, in the spectra we have information from outside the visible range that we can use to distinguish leaf types. This is most apparent in cluster 4 where the differences in the spectra outside the visible range are most stark.

When we look at just the visible range for the reflectance in Fig. 4.21 we see that the 3 of clusters have similar shapes below 700 nm. These clusters, clusters 1, 2, and 4, have a green color in Fig. 4.20. If we only had data from the visible range, our analysis may not have found as many clusters as we found from using the hyper-spectral data.

SVD and K-means Combined

In this case study combining an SVD and K-means gave us further insight into the data. By using a four-term SVD approximation to the data matrix we were able to classify the leaves into 5 types, including one that could not be distinguished looking solely at the visible spectrum.

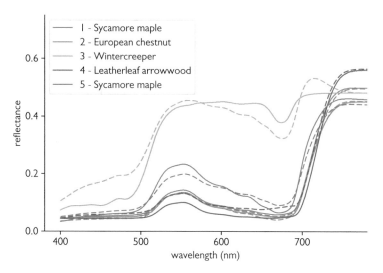

Fig. 4.21 Comparison of cluster centers (dashed lines) and nearest leaf in the data set for reflectance in the visible range

Notes

Additional discussion of the SVD and its properties can be found in [5].

We have only covered a small subset of unsupervised learning techniques. We will return to this problem later when we discuss autoencoders. There are variants of the SVD that have different features. These include independent components [6], factor analysis [7], and non-negative matrix factorization [8]. Non-negative factorization is especially interesting in applications where the variables are known to be positive (as occurs often in science and engineering). As mentioned above, there are variants of K-means clustering that are useful. Descriptions of many of these methods can be found in [9].

Problems

4.1 Generate 100 time series from the function

$$y(t) = w_1 \sin(2\pi t) + w_2 \cos(2\pi t) + \epsilon_t$$

for $t = 0, 0.05, 0.1, \ldots, 1.0$ and w_1 and w_2 are random numbers between 0 and 1 that are constant for each time series and ϵ_t is a sample from a normal distribution with mean 0 and standard deviation 0.05 at each time point. Using this data, create a data matrix, perform an SVD, and see what rank you need to describe 95% of the data.

4.2 The iris classification data set is available in scikit-learn. Use K-means and t-SNE to cluster this data. There are three classes in the data. For K-means with $k = 3$ do the clusters correspond to the 3 classes. Can you see the three clusters in the t-SNE results?

4.3 The Olivetti face data set of 400 grayscale images of 64×64 pixels is available in scikit-learn. Each image is the face of one of 40 people. Consider each face as a vector of length 4096 to create a data matrix of size 400×4096. Perform an SVD on this data matrix and visualize the first 5 columns as images. What do these images tell you about the faces in the data set? Now perform a K-means clustering of the faces data set with $k = 40$. Does K-means find the forty different people? Finally, apply t-SNE to the data and see what groupings it finds in the data.

References

1. Laurens van der Maaten and Geoffrey Hinton. Visualizing Data using t-SNE. *Journal of Machine Learning Research*, 9(1):2579–2605, November 2008.
2. Han Xiao, Kashif Rasul, and Roland Vollgraf. Fashion-MNIST: a novel image dataset for benchmarking machine learning algorithms. *arXiv preprint arXiv:1708.07747*, 2017.
3. S Jacquemound, L Bidel, C Francois, and G Pavan. ANGERS leaf optical properties database (2003). *Data Set. Available online: http://ecosis.org (accessed on 19 June 2019)*, 2003.
4. T Smith and J Guild. The C.I.E. colorimetric standards and their use. *Transactions of the Optical Society*, 33(3):73–134, Jan 1931.
5. Ryan G McClarren. *Uncertainty Quantification and Predictive Computational Science*. Springer, 2018.
6. J.V. Stone and Massachusetts Institute of Technology. MIT. *Independent Component Analysis: A Tutorial Introduction*. A Bradford Book. MIT Press, 2004.
7. An Gie Yong and Sean Pearce. A beginner's guide to factor analysis: Focusing on exploratory factor analysis. *Tutorials in quantitative methods for psychology*, 9(2):79–94, 2013.
8. Trevor Hastie, Robert Tibshirani, and Jerome Friedman. *The Elements of Statistical Learning*. Data Mining, Inference, and Prediction, Second Edition. Springer Science and Business Media, New York, NY, August 2009.
9. Anil K Jain, Richard C Dubes, et al. *Algorithms for clustering data*, volume 6. Prentice Hall, 1988.

Part II
Neural Networks

Chapter 5
Feed-Forward Neural Networks

Non omnia lusibus data sunt, sed nonnulla conflictibus[1]

—*Joachim of Fiore from* Liber de Concordia Noui ac
Veteris Testamenti

Abstract In this chapter we develop the theory and applications of feed-forward neural networks for supervised learning problems. We begin by starting with the linear regression models previously developed and add a nonlinear transformation to create a neural network with a single hidden layer. We discuss how these networks are trained via stochastic gradient descent. Then we discuss how deep neural networks are formed by adding further hidden layers to the network. The topics of regularization and dropout are included to give the reader the ability to use these sophisticated (and useful) tools. We conclude with a case study where deep neural networks are used to predict the strength of concrete given several independent variables. This chapter is the basis for most of the remaining chapters where neural networks will be used in the subsequent discussions.

Keywords Neural networks · Hidden layers · Activation functions · Training neural networks · Data normalization · Stochastic gradient descent · Deep neural networks · Regularization; Dropout

5.1 Simple Neural Network

In Chap. 2 we explored linear models to develop a relationship between independent and dependent variables. These models took the form of a linear combination of the independent variables plus a bias term (sometimes called an intercept). For a dependent variable y, and J independent variables x_j, a linear model takes the form:

---

[1]Data do not dance and play; there are some conflicts.

© Springer Nature Switzerland AG 2021
R. G. McClarren, *Machine Learning for Engineers*,
https://doi.org/10.1007/978-3-030-70388-2_5

$$y = w_1 x_1 + w_2 x_2 + \cdots + w_J x_J + b.$$

In a neural network we introduce a nonlinear transformation on top of the linear combination. To introduce this form we create a series of intermediate variables, z_k for $k = 1, \ldots, K$, that are functions of a linear combination of the inputs. Specifically, we write

$$z_k = \sigma(w_{k1} x_1 + w_{k2} x_2 + \cdots + w_{kJ} x_J + b_k). \tag{5.1}$$

In this definition of the intermediate variable we have used a function $\sigma(u)$ called an activation function or a sigmoid function. There are a variety of forms for the activation function that we will detail later.

The dependent variable is then a function of a linear combination of the z_k:

$$y = o(w_{o1} z_1 + w_{o2} z_2 + \cdots + w_{oK} z_K + b_o), \tag{5.2}$$

where o is the output function.

A schematic of this simple neural network is shown in Fig. 5.1. In the figure the input data is fed into the network at the left and the inputs are combined via a linear combination with the bias added and to form the intermediate variables z_k. These intermediate variables together are often called a hidden layer, because these intermediate values are in between the inputs and the output and are typically

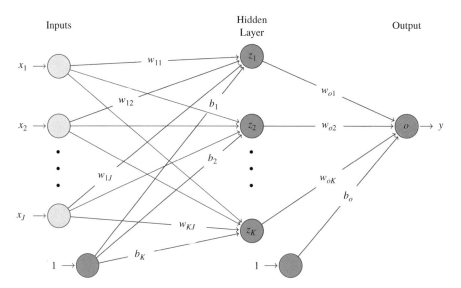

Fig. 5.1 Schematic of a simple feed-forward neural network defined by Eqs. (5.1) and (5.2) with information flowing from the left to right. The inputs x_j are multiplied by weights, added together with the bias, and passed to the function σ to get the intermediate values z_k, also known as hidden units. The z_k are then combined with a bias and fed to the output function o to get the value of y

not seen by the user of the neural network. The individual z_k are sometimes called hidden units, nodes, or neurons. The hidden units are then combined via a linear combination with the bias added to produce the output after being passed to an output function, o.

The particular network in Fig. 5.1 is sometimes called a feed-forward neural network because information travels only in one direction, left to right in this case: information is fed on the left and that information flows to the hidden layer before flowing to the output. This contrasts with recurring neural networks and other forms where information might travel in loops.

Feed-Forward Neural Network with a Single Hidden Layer
A neural network with a single hidden layer looks very similar to the linear regression models we previously discussed with added nonlinearity through the activation function and the output function. The network is called "feed-forward" because the information travels in a single direction from the inputs to the outputs.

5.1.1 Example Neural Network

One nonlinear function that we can model with this simple neural network is the exclusive-OR function (XOR). This function takes two binary inputs that are 0 or 1 and returns 1 only if one of the inputs is 1, and returns 0 if both inputs are 0 or 1:

$$y(x_1, x_2) = \begin{cases} 1 & x_1 + x_2 = 1 \\ 0 & \text{otherwise} \end{cases}, \qquad x_1, x_2 \in \{0, 1\}. \tag{5.3}$$

We can exactly fit this function with the simple neural network if we define

$$z_1 = \sigma_S(x_1 - 1), \tag{5.4a}$$

$$z_2 = \sigma_S(x_1 + x_2 - 2), \tag{5.4b}$$

$$z_3 = \sigma_S(x_2 - 1), \tag{5.4c}$$

where the activation function is the step function:

$$\sigma_S(u) = \begin{cases} 1 & u \geq 0 \\ 0 & \text{otherwise} \end{cases}. \tag{5.5}$$

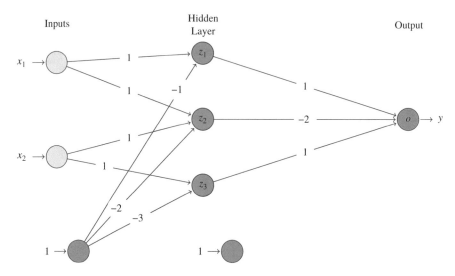

Fig. 5.2 Visualization of the neural network that reproduces the XOR function. The activation function used in the hidden layer is the step function σ_S

Then for y we define

$$y = z_1 - 2z_2 + z_3 \tag{5.6}$$
$$= \sigma_S(x_1 - 1) - 2\sigma_S(x_1 + x_2 - 2) + \sigma_S(x_2 - 1).$$

In this case the output function is the identity.

This neural network is drawn in Fig. 5.2. All connections that are zero, i.e., zero weights or biases, are not drawn. In this network we see that x_1 is only connected to z_1 and z_2 and that x_2 is only connected to z_2 and z_3. Furthermore, the output bias is 0 so there is no connection between the output bias and the output node.

We can check that Eqs. (5.4) and (5.6) reproduce the XOR function. We first test that $x_1 = x_2 = 1$ gives $y = 0$:

$$z_1 = \sigma_S(1 - 1) = 1, \quad z_2 = \sigma_S(1 + 1 - 2) = 1, \quad z_3 = \sigma_S(1 - 1) = 1,$$

$$y = 1 - 2 + 1 = 0.$$

Also, if $x_1 = 1$ and $x_2 = 0$ the function should output $y = 1$:

$$z_1 = \sigma_S(1 - 1) = 1, \quad z_2 = \sigma_S(1 + 0 - 2) = 0, \quad z_3 = \sigma_S(0 - 1) = 0,$$

$$y = 1 - 0 + 0 = 1.$$

We can see that if $x_1 = 0$ and $x_2 = 1$ the function will also return $y = 1$. Finally, $x_1 = x_2 = 0$ gives $y = 0$:

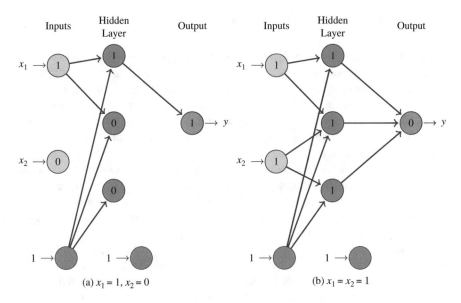

Fig. 5.3 The flow of information through the neural network for the cases: (**a**) $x_1 = 1, x_2 = 0$ and (**b**) $x_1 = x_2 = 1$. Only connections that are non-zero are drawn for each case. We observe that the hidden layer shuts off information transfer depending on the inputs

$$z_1 = \sigma_S(0 - 1) = 0, \quad z_2 = \sigma_S(0 + 0 - 2) = 0, \quad z_3 = \sigma_S(0 - 1) = 0,$$

$$y = 0 - 0 + 0 = 0.$$

One feature of neural networks is that the activation functions in the hidden layer control how information flows through the network from the inputs to the output. In particular we can see how the cases $x_1 = 1, x_2 = 0$ and $x_1 = x_2 = 1$ have information flow through the network in Fig. 5.3. In the figure we see that in the case of $x_1 = 1, x_2 = 0$ the activation functions in z_2 and z_3 "shut off" and propagate no further information. However, when $x_1 = x_2 = 1$ all of the hidden layer functions flow information to the output. The weights of the connections to the output function assure that the information that does flow to the output node is combined appropriately to the give the correct value.

Nonlinearity in the Activation Function
The activation function allows the network to have different pathways through it by shutting of, or setting to zero, certain values in the hidden layer. This type of nonlinearity is required to reproduce a nonlinear function like XOR

5.1.2   Why Is This Called a Neural Network?

To this point we have seen that feed-forward networks have the property that information flows from the inputs to the outputs and that the activation functions act to shut off information flow through some channels. We have not stated why this might be called a "neural" network. The primary reason for this is that the units in the hidden layer, the intermediate values we call z_k, act in a very similar way to the biological neurons found in the brain. A neuron is a cell that has several inputs— called dendrites—that receive chemo-electrical signals. When a large enough signal is received at the dendrites, the neuron activates, or, more colloquially, fires, by sending an electrical signal through an output channel called an axon. The axon can be connected to other neurons and the firing of the axon can cause other neurons to fire. The signal required for the neuron to fire can be affected by chemicals in the brain, called neurotransmitters.

The structure of a neural network resembles the biological neurons. Loosely speaking, the dendrites are the inputs to a node in the hidden layer, the bias are the neurotransmitters, the activation function, σ, shuts off the signal unless a threshold is reached, and the outputs of the hidden layer are the axons. The feed-forward network we have considered is only a crude approximation of neural structure, but the analogy can be helpful to think about how the network functions.

5.1.3   The Activation Function, $\sigma(u)$

The activation function, $\sigma(u)$, typically has the property that for u below a certain value, the function returns 0. We saw this above with the step function $\sigma_S(u)$: when $u < 0$ the function returns 0. The activation function then acts to shut off the flow of information through the network when the signal received is less than some threshold, and this threshold is controlled by the bias.

The step function is one of several possibilities of the activation function. Another function that is commonly used is one that we have already seen, the logistic function:

$$\sigma_L(u) = \frac{1}{1 + e^{-u}}. \tag{5.7}$$

Notice that this function has the same limits as the step function: as $u \to \infty$, $\sigma_L(u) \to 1$, and $u \to -\infty$, $\sigma_L(u) \to 0$. One difference between the logistic function and the step function is that the logistic function is smooth: its derivative exists everywhere and is positive everywhere. This is not true for the step function where the derivative is zero everywhere except at $u = 0$ where the derivative is undefined. The logistic function is plotted in Fig. 5.4.

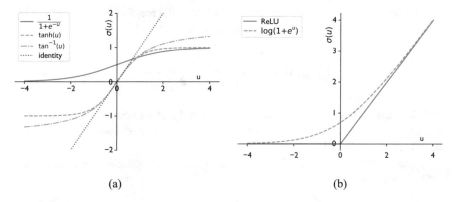

Fig. 5.4 Comparison of several activation functions. (**a**) The functions that have characteristic "s" shape, the logistic, hyperbolic tangent, and arctangent are compared. The identity function is shown for comparison near the origin. (**b**) The ReLU and softplus functions have similar behavior, but the softplus function is smooth

There are several other functions that have similar behavior to the logistic function. One function is the hyperbolic tangent function, $\tanh(u)$. This function has a similar shape to the logistic function but has a range from -1 to 1. We denote an activation function using the hyperbolic tangent as $\sigma_{TH}(u)$:

$$\sigma_{TH}(u) = \tanh(u). \tag{5.8}$$

Though this function does not "shut off" by outputting zero when the input is low enough, it has the property that if the input is much less than zero, the function returns a known value of -1. In this way, the information can be shut off if we add 1 to the output of the function.

Additionally, the arctangent function can be an activation function:

$$\sigma_{AT}(u) = \tan^{-1}(u). \tag{5.9}$$

The hyperbolic tangent and arctangent activation functions are smooth, like the logistic function, and they have the additional property that they behave like the identity function near the origin. At $u = 0$ both $\sigma_{TH}(0) = 0$ and $\sigma_{AT}(0) = 0$. Also, the derivative of these functions with respect to u is 1 at the origin. This makes the function locally similar to the identity. This will have important applications when we train the networks later. The similarity to the identity can be seen in Fig. 5.4 where the hyperbolic tangent and arctangent are compared to the identity.

In Fig. 5.4 we notice that the limits of the arctangent and the hyperbolic tangent functions are not the same as the logistic function (or the step function). Nevertheless, one could shift and scale these functions so that they had the same range as the logistic function.

Another important type of activation function is the rectified linear unit (ReLU) function that returns 0 for $u \leq 0$ and u for $u > 0$:

$$\sigma_{\text{ReLU}}(u) = \begin{cases} 0 & u \leq 0 \\ u & u > 0 \end{cases} = \frac{1}{2}\left(|u| + u\right). \tag{5.10}$$

This function "shuts off" when u is negative and can increase without bound as u increases above 0. This function has a derivative defined everywhere, but the derivative jumps from 0 to 1 at $u = 0$.

There is also a smooth version of the ReLU called the softplus that smooths the transition to zero:

$$\sigma_{\text{SP}}(u) = \log\left(1 + e^u\right). \tag{5.11}$$

This function behaves similar to the identity function as $u \gg 0$ because $\sigma_{\text{SP}}(u) \approx \log(e^u) = u$ for u large, and approaches 0 as $u \to -\infty$. The derivative of the softplus function is the logistic function, and the derivative is therefore well-defined and positive everywhere.

The ReLU and softplus activation functions are shown in Fig. 5.4b. In the figure it is clear that as $u \to \pm\infty$ the value of the softmax function approaches that of the ReLU function; near the origin the softplus function is smooth.

Another function that we can use in the role of an activation function is the softmax function that was discussed in Sect. 2.3.4. In this case the function takes in J values and outputs J values. These will be used when we consider classification models with neural networks.

Activation Functions

There are a variety of functions that we can use for the activation of a hidden unit in a neural network. Most of them mimic a biological neuron by having the ability to return a value of zero below a certain value. This is how the network can control which units have a non-zero value based on the inputs to the unit.

5.2 Training Single Layer Neural Networks

To understand how we determine the weights and biases of the neural network we will consider a single layer network, like that shown in Fig. 5.1. We consider a training set of data consisting of a single pair of independent variables, $\mathbf{x}$, and dependent variable y. We want to minimize the error between the neural network prediction $\hat{y}$ and the true value y. We use the squared error as our loss function:

$$L = (\hat{y} - y)^2. \tag{5.12}$$

To minimize this loss function we want to compute the derivative of the loss function with respect to each of the parameters, i.e., all the weights and biases. Then we want to adjust each parameter to make the loss function smaller in magnitude. For instance, if we have the derivative of the loss function with respect to a given weight, w_{kj}, we can use the gradient descent method to decrease the loss function by updating the weight as

$$w_{kj}^{\text{new}} = w_{kj} - \eta \frac{\partial L}{\partial w_{kj}}, \tag{5.13}$$

where $\eta > 0$ is called the learning rate and governs how large the magnitude of the change to the weight needs to be. In the gradient descent method we repeatedly update the weights until we find a minimum value of the loss function. This is called gradient descent because we adjust the weights in the opposite direction of the gradient.

In what follows we present how to compute the derivative of the loss function with respect to the weights and biases. We repeat the equations for the kth value in the hidden layer, z_k, and the equation for $\hat{y}$

$$z_k = \sigma(w_{k1}x_1 + w_{k2}x_2 + \cdots + w_{kJ}x_J + b_k) = \sigma(u_k), \tag{5.14}$$

$$\hat{y} = o(w_{o1}z_1 + w_{o2}z_2 + \cdots + w_{oK}z_K + b_o) = o(u_o). \tag{5.15}$$

In these equations we have written the sums that form the inputs to the activation function and the output function using $u_k = w_{k1}x_1 + w_{k2}x_2 + \cdots + w_{kJ}x_J + b_k$, and $u_o = w_{o1}z_1 + w_{o2}z_2 + \cdots + w_{oK}z_K + b_o$. This will make it clear that these functions only take a single input for the purposes of taking derivatives.

Now we compute the derivative of the loss function with respect to one of the weights in the output layer, w_{ok},

$$\frac{\partial L}{\partial w_{ok}} = (\hat{y} - y)\frac{\partial \hat{y}}{\partial w_{ok}} \tag{5.16}$$

$$= (\hat{y} - y)\frac{do}{du_o}\frac{\partial u_o}{\partial w_{ok}}$$

$$= (\hat{y} - y)\frac{do}{du_o}z_k$$

$$= \delta_o z_k.$$

Here we have defined a quantity δ_o as the product of the derivative of the output function and the difference of $\hat{y}$ and y:

$$\delta_o = (\hat{y} - y)\frac{do}{du_o}.$$

Similarly, if we consider the derivative with respect to the bias in the output layer, we have

$$\frac{\partial L}{\partial b_o} = (\hat{y} - y)\frac{\partial \hat{y}}{\partial b_o} \tag{5.17}$$

$$= (\hat{y} - y)\frac{do}{du_o}\frac{\partial u_o}{\partial b_o}$$

$$= \delta_o.$$

Notice that the derivative of the loss function with respect to the bias and the weights in the output layer depend on the difference between the output and the true value of the dependent variable times the derivative of the output function. In the case of the weights, there is the additional term containing the value of the hidden layer. This makes the derivative of the loss function with respect to the weights easy to evaluate provided that we can compute the derivative of the output function. Also, notice that we only have to compute δ_o once and can use it to evaluate all $K+1$ derivatives we need.

We can do the same calculation for the weights and biases that form the hidden layer values, z_k. The derivative with respect to w_{kj} is

$$\frac{\partial L}{\partial w_{kj}} = (\hat{y} - y)\frac{\partial \hat{y}}{\partial w_{kj}} \tag{5.18}$$

$$= (\hat{y} - y)\frac{do}{du_o}\frac{\partial u_o}{\partial z_k}\frac{\partial z_k}{\partial w_{kj}}$$

$$= \delta_o w_{ok}\frac{d\sigma}{du_k}\frac{\partial u_k}{\partial w_{kj}}$$

$$= \delta_o w_{ok}\frac{d\sigma}{du_k}x_j$$

$$= \delta_k x_j.$$

Here we have written

$$\delta_k = \delta_o w_{ok}\frac{d\sigma}{du_k}.$$

This allows us to notice that the derivative with respect to a weight feeding the hidden layer can be expressed in terms of a product of the derivative of the output function, the derivative of the activation function, and the weight connecting the hidden layer to the output layer.

The derivative with respect to the bias b_k can be computed in a similar fashion:

$$\frac{\partial L}{\partial b_k} = (\hat{y} - y)\frac{\partial \hat{y}}{\partial b_k} \qquad (5.19)$$

$$= (\hat{y} - y)\frac{do}{du_o}\frac{\partial u_o}{\partial z_k}\frac{\partial z_k}{\partial b_k}$$

$$= \delta_o w_{ok}\frac{d\sigma}{du_k}\frac{\partial u_k}{\partial w_k}$$

$$= \delta_o w_{ok}\frac{d\sigma}{du_k}$$

$$= \delta_k.$$

This procedure for calculating the derivative of the loss function with respect to the weights/biases by starting at the output and working toward the inputs is called back propagation. This is because we first compute the term δ_o for the output layer and then use it to form δ_k for the hidden layer.

5.2.1 Multiple Training Points

Above we considered only a single training point in computing the derivative of the loss function with respect to weights/biases. If we have I training points, we can write the input data as a matrix $\mathbf{X}$ of size $I \times J$ with each row containing the J inputs for case i. This also implicitly defines a matrix $\mathbf{Z}$ of size $I \times K$ where each row contains the K values for z_k for case i. Using this notation, we then write the loss function as

$$L = \sum_{i=1}^{I}(\hat{y}_i - y_i)^2. \qquad (5.20)$$

A straightforward calculation reveals that the derivatives in this case can be expressed in terms of

$$\delta_{oi} = (\hat{y}_i - y_i)\frac{do}{du_o}, \qquad \delta_{ki} = \delta_{oi} w_{ok}\frac{d\sigma}{du_k}, \qquad (5.21)$$

as

$$\frac{\partial L}{\partial w_{ok}} = \sum_{i=1}^{I}\delta_{oi} z_{ik} \qquad \frac{\partial L}{\partial b_o} = \sum_{i=1}^{I}\delta_{oi} \qquad (5.22)$$

$$\frac{\partial L}{\partial w_{kj}} = \sum_{i=1}^{I} \delta_{ki} x_{ij} \qquad \frac{\partial L}{\partial b_k} = \sum_{i=1}^{I} \delta_{ki}.$$

These equations tell us that we can compute the derivative of the loss function with respect to the loss function over a training set by adding the derivatives from each training point.

> **Back Propagation to Compute the Gradient of the Loss Function**
> By starting at the output layer, we can compute the derivative of the loss function with respect to the weights/biases in the network. We need this gradient to update the weights via gradient descent. The calculation involves the derivative of the activation and output functions, and the values entering a unit.

5.2.2   Data Normalization and Training Neural Networks

At this point it is a good time to discuss one consequence of the nonlinear transformations that arise from the activation functions and the gradient computation we just performed: the necessity for normalization of network inputs and outputs. As discussed above the activation functions have specific behavior near $u = 0$, namely that many of them have the behavior that near $u = 0$ they behave somewhat like the identity function. This makes

$$\frac{d\sigma}{du} \approx 1,$$

for $u \approx 0$. Also, in Eq. (5.22) we see that the gradient of the loss function contains a term that is multiplied by the input value x_{ij}. If the x_{ij} values vary wildly in magnitude between the independent variables (i.e., the range of the x_{ij} is much larger than the range of $x_{ij'}$), the weights affected will have a large variation in magnitude in their gradients. Additionally, if within a single independent variable there is a large range between cases i and i', the sum in Eq. (5.22) will be adding contributions of very different scales (and perhaps signs). Furthermore, all of the terms in Eqs. (5.21) and (5.22) have $(\hat{y}_i - y_i)$ in them. If the outputs have a large variation to them, the magnitude of this term will have a large variation. Combined will cause the contribution from each training point and independent variable to have very different effects on the gradient computation.

While it is possible to apply gradient descent to problems where the gradients have large variations in magnitude and sign, it will train much faster when the contribution from each independent variable is on the same scale *and* when the outputs are on a bounded or nearly bounded scale. For this reason it is the standard practice to normalize the independent variables so that each has a fixed range and variability. For instance, we can subtract from each variable its mean and then divide by its standard deviation (this is sometimes called computing a z-score in statistics). In this process we produce a new variable $\tilde{x}_{ij}$ as

$$\tilde{x}_{ij} = \frac{x_{ij} - \bar{x}_j}{s_j}, \tag{5.23}$$

where $\bar{x}_j$ is the mean of the jth independent variable and s_j is the sample standard deviation of this variable. The normalization constants $\bar{x}_j$ and s_j must be stored so that new data and validation data can be placed on the same scale. Furthermore, we can apply the same transform to the output data y to normalize the dependent variables. After applying the normalization the data will have mean 0 and standard deviation of 1. In other words the data have a uniform variability between the variables and all variables have the value of 0 correspond to the mean value of that variable.

This type of normalization of the inputs and outputs of a neural network is the standard practice and makes training the networks much easier in practice. We use such normalizations (or variants when noted) throughout this work when dealing with neural networks.

Data Normalization for Neural Networks
One should *always* normalize the input and output data when training a neural network. Due to the nonlinearity in neural networks, normalization makes the training converge much more rapidly and is the standard practice in machine learning. A standard normalization is to subtract the mean of the variable and divide by its standard deviation.

5.2.2.1 Stochastic Gradient Descent

A modification of the standard gradient descent method is stochastic gradient descent (SGD). SGD is a modification of gradient descent where the gradient of the loss function with respect to the weights is estimated on a subset of the available training data. This subset is called a batch and the process of training on each batch is called an epoch. If we have I training points, and we divide into b batches of size I_b, we then have a procedure to iterate through the batches, updating the weights for each batch, and then shuffling the training points into different batches. The algorithm looks like the following:[2]

- Repeat until the loss function is small enough:
  - Shuffle training points between batches randomly.[3]
  - For each of the b batches:

    Evaluate Eq. (5.22) using gradients estimated over the I_b points in the batch.
    Update the weights using Eq. (5.13).

A beneficial feature of stochastic gradient descent is that it works well if more training points become available while we are training the model: the new points are just more data to shuffle between the batches. Additionally, if we have a large number of training points we do not have to use all of the training points. We can randomly select a subset of them in each iteration and shuffle them between batches. This will give us a representative way to update models without evaluating the loss function gradients at every training point available.

Stochastic Gradient Descent (SGD)
Stochastic gradient descent (SGD) is a method for applying the gradient descent algorithm to update the weights and biases in a network using only a subset of the available training data, called a batch. Each batch provides an update to the weights and biases of the model. This approach is useful when the training data is large or when new data becomes available during training.

5.2.2.2   Issues with Gradient Descent

Above we mentioned that we can iteratively update the weights and biases by adjusting them in the direction opposite to the derivative of the loss function. This requires derivatives to be calculable and defined, which eliminates some choices for activation functions, such as step functions, unless the derivative is approximated. Another, more vexing, issue with the gradient descent approach is that because each update decreases the loss function, it is susceptible to finding a *local* minimum rather than the global minimum. This can be seen in the example shown in Fig. 5.5. In the

---

[2]Sometimes this method is called mini-batch stochastic gradient descent, with SGD reserved for the case where the batch size is 1. Given the ubiquity of the algorithm, dropping the "mini-batch" descriptor should not lead to any confusion.

[3]It is possible to skip this shuffling step, but in general shuffling the batches assures that the order the batches are processed does not matter.

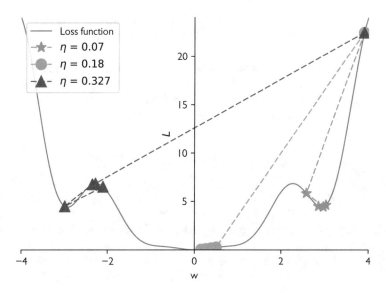

Fig. 5.5 Demonstration of four iterations of the gradient descent finding a local minimum: the smooth curve is the loss function and we have an initial value of the parameter w at the top right. The value is updated by gradient descent, as shown by the symbols. Different magnitudes of the learning rate, η, lead to different minima found

figure we have an example loss function that is the function of a single parameter, w. If the initial value is $w = 3.9$ and we update w using gradient descent, then the minimum found depends on the learning rate used. It is also possible that if the learning rate is too large the method will not converge.

Given that the minimum found is sensitive to the learning rate in this simple, one parameter optimization problem, it is not hard to imagine that in applying gradient descent to a large number of weights and biases that local minima are likely to be found. For this reason there is much active research in finding the weights and biases that minimize the loss function. Most software libraries for building neural networks have multiple optimization options beyond simple gradient descent (or its stochastic version). These methods are outside the scope of the discussion, but it is important to know that a neural net is only as good as the optimization solution found. Some of these methods use the concept of momentum to help the stochastic gradient descent converge. The idea is to make the new update a linear combination the previous change to the weights plus the estimated change from the gradient:

$$w_{kj}^{\text{new}} = w_{kj} - \eta \frac{\partial L}{\partial w_{kj}} + \alpha \Delta w_{kj},$$

where Δw_{kj} is the change in w_{kj} from the previous iteration, and $\alpha > 0$. In this case we can think of the weights as having their update depending on the most recent update, as though it is an object traveling with momentum. The weights move along their update path and that path is adjusted by the "force" represented by the gradient of the loss function.

Another technique that has been found to be effective and simple is to start with a relatively large learning rate and decrease it as the number of iterations increases. The idea here is to try and skip over local minima in the early iterations and then zoom in on the global minimum in the later iterations.

Issues with Gradient Descent

Gradient descent and its stochastic version are susceptible to finding weights and biases that are only a local minimum for the loss function. This means the neural network model that the training converges to does not have the weight and bias parameters that give the lowest error. There are modifications to the gradient descent method that attempt to address these shortcomings, but they are typically variations on the idea of gradient descent.

5.3 Deep Neural Networks

The real power of neural networks arises when the network has more than a single hidden layer. Neural networks with multiple hidden layers are called deep neural networks (DNN). These networks are an extension of the simple networks we have studied earlier in this chapter. In DNNs there are several intermediate values rather than a single set. A neural network with 3 hidden layers is shown in Fig. 5.6. This network has K hidden units per hidden layer. Notice that as we add layers the number of connections between nodes grows, making the number of weights and biases that we must find grow as well. Note that this is still a feed-forward network in that information is only moving from left to right in the figure: there are no connections between units in the same layer or going backward from right to left.

To describe a deep neural network we need to define some notation for the weights and biases. The number of hidden layers is L. We define a matrix $\mathbf{A}^{\ell \to \ell'}$ to be the matrix containing the weights and the bias that connect layer ℓ to layer ℓ'. We also write that layer ℓ has $K^{(\ell)}$ weights and layer ℓ' has $K^{(\ell')}$ weights; this matrix has the size $(K^{(\ell'+1)} \times K^{(\ell+1)}$, and the additional 1 is for the bias. We write the vector of hidden units in layer ℓ as $\mathbf{z}^{(\ell)} = (z_1^{(\ell)}, z_2^{(\ell)}, \ldots, z_{K^{(\ell)}}^{(\ell)}, 1)^{\mathrm{T}}$ including the 1 for the bias. To make the notation general, we consider the input layer as layer number 0 of length $K^{(0)} = J$ and $\mathbf{z}^{(0)} = (x_1, x_2, \ldots, x_J, 1)^{\mathrm{T}}$. The output layer is considered layer $L + 1$ of size $K^{(L+1)} = 1$. The matrix $\mathbf{A}^{\ell \to \ell'}$ will have the form

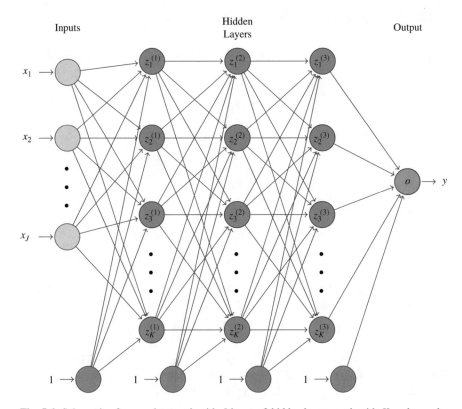

Fig. 5.6 Schematic of a neural network with J inputs, 3 hidden layers, each with K nodes, and a single output

$$\mathbf{A}^{\ell \to \ell'} = \begin{pmatrix} w_{11}^{\ell \to \ell'} & w_{12}^{\ell \to \ell'} & \ldots w_{1\ell}^{\ell \to \ell'} & b_1^{\ell \to \ell'} \\[2mm] w_{21}^{\ell \to \ell'} & w_{22}^{\ell \to \ell'} & \ldots w_{2\ell}^{\ell \to \ell'} & b_2^{\ell \to \ell'} \\[2mm] \vdots & & & \\[2mm] w_{K^{(\ell')}1}^{\ell \to \ell'} & w_{K^{(\ell')}2}^{\ell \to \ell'} & \ldots w_{K^{(\ell')}\ell}^{\ell \to \ell'} & b_{K^{(\ell')}}^{\ell \to \ell'} \\[2mm] 0 & \ldots & & 1. \end{pmatrix} \qquad (5.24)$$

In the matrix we have written the weight that connects node j from layer ℓ to node i in layer ℓ' as $w_{ij}^{\ell \to \ell'}$ and the bias that feeds node i in layer ℓ' as $b_i^{\ell \to \ell'}$. Note the matrix connecting the last hidden layer to the output, $\mathbf{A}^{(L \to L+1)}$, will not have the final row.

Using our notation we write the network depicted in Fig. 5.6 as

$$\mathbf{z}^{(1)} = \sigma\left(\mathbf{A}^{0\to 1}\mathbf{z}^{(0)}\right),$$ (5.25a)

$$\mathbf{z}^{(2)} = \sigma\left(\mathbf{A}^{1\to 2}\mathbf{z}^{(1)}\right),$$ (5.25b)
$$= \sigma\left(\mathbf{A}^{1\to 2}\sigma\left(\mathbf{A}^{0\to 1}\mathbf{z}^{(0)}\right)\right),$$

$$\mathbf{z}^{(3)} = \sigma\left(\mathbf{A}^{2\to 3}\mathbf{z}^{(2)}\right),$$ (5.25c)
$$= \sigma\left(\mathbf{A}^{2\to 3}\sigma\left(\mathbf{A}^{1\to 2}\sigma\left(\mathbf{A}^{0\to 1}\mathbf{z}^{(0)}\right)\right)\right),$$

$$y = o\left(\mathbf{A}^{3\to 4}\mathbf{z}^{(3)}\right),$$ (5.25d)
$$= o\left(\mathbf{A}^{3\to 4}\left(\sigma\left(\mathbf{A}^{2\to 3}\sigma\left(\mathbf{A}^{1\to 2}\sigma\left(\mathbf{A}^{0\to 1}\mathbf{z}^{(0)}\right)\right)\right)\right)\right).$$

These equations show how the output y is a function of the inputs. The inputs go through several linear combinations and several nonlinear functions. We will now demonstrate one of these networks on a simple example.

Deep Neural Networks
Adding hidden layers to a feed-forward neural network creates a deep neural network (DNN). These networks are more flexible than neural networks with a single hidden layer because there are multiple nested linear combinations and nonlinear transformations performed to the input data.

5.3.1   Example Deep Neural Network

Consider the function $\sin(x)$ for $x \in [0, 2\pi]$. A neural network to approximate this function is displayed in Fig. 5.7; this network uses ReLU for the activation functions and the output function is the identity. Using our notation the weight/bias matrices are

Fig. 5.7 A neural network to approximate the sine function. The connections drawn represent the non-zero weights/biases. The hidden units use the ReLU activation function and the output function o is the identity

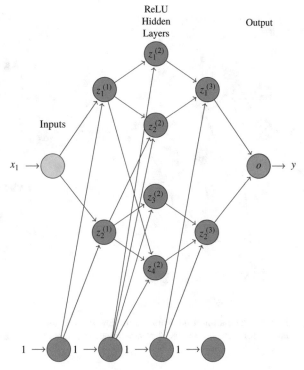

$$\mathbf{A}^{0\to1} = \begin{pmatrix} 1 & -\pi \\ -1 & \pi \\ 0 & 1 \end{pmatrix}, \quad \mathbf{A}^{1\to2} = \begin{pmatrix} 1 & 0 & -\frac{\pi}{2} \\ -1 & -10^6 & \frac{\pi}{2} \\ 0 & 1 & -\frac{\pi}{2} \\ -10^6 & -1 & \frac{\pi}{2} \\ 0 & 0 & 1 \end{pmatrix}, \quad (5.26)$$

$$\mathbf{A}^{2\to3} = \begin{pmatrix} -0.664439 & -0.664439 & -10^6 & -10^6 & 1.15847 \\ -10^6 & -10^6 & -0.664439 & -0.664439 & 1.15847 \\ 0 & 0 & 0 & 1 \end{pmatrix},$$

$$\mathbf{A}^{3\to4} = \begin{pmatrix} -1 & 1 & 0 \end{pmatrix}.$$

To see how this network works, we will look at two potential inputs. If $x_1 = \pi/4$, as shown in Fig. 5.8a, the ReLU activation functions force information to flow only along the path that goes from the input to the output as $x_1 \to z_2^{(1)} \to z_3^{(2)} \to z_2^{(3)} \to o$. Using the definition of the weight matrices, we get the hidden layers' values as

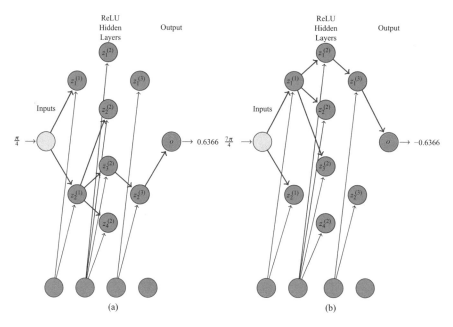

Fig. 5.8 Illustration of the information flow through the neural network to approximate the sine function for different inputs. The only connections that are non-zero are drawn. (**a**) $x_1 = \pi/4$. (**b**) $[x_1 = 7\pi/4]$

$$\mathbf{z}^{(1)} = \begin{pmatrix} 0 \\ 2.3561 \\ 1 \end{pmatrix}, \quad \mathbf{z}^{(2)} = \begin{pmatrix} 0 \\ 0 \\ 0.78539 \\ 0 \\ 1 \end{pmatrix}, \quad \mathbf{z}^{(3)} = \begin{pmatrix} 0 \\ 0.63662 \\ 1 \end{pmatrix}, \quad y = 0.63662.$$

$$(5.27)$$

If the input is $x_1 = 7\pi/4$, the path through the network is different, as illustrated in Fig. 5.8b. Here information flows along the topmost path of the network. In this case the hidden layers are

$$\mathbf{z}^{(1)} = \begin{pmatrix} 2.3561 \\ 0 \\ 1 \end{pmatrix}, \quad \mathbf{z}^{(2)} = \begin{pmatrix} 0.78539 \\ 0 \\ 0 \\ 0 \\ 1 \end{pmatrix}, \quad \mathbf{z}^{(3)} = \begin{pmatrix} 0.63662 \\ 0 \\ 1 \end{pmatrix}, \quad y = 0.63662.$$

$$(5.28)$$

Notice that this function preserves the property of the sine function that $\sin x = -\sin(x + \pi)$.

Going beyond these two example inputs, we point out what this network is really doing: it is using a piecewise linear approximation to the sine function using four

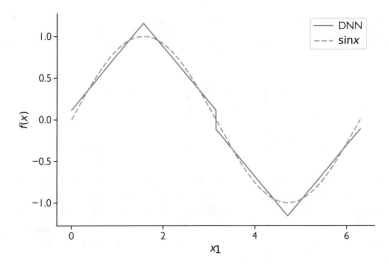

Fig. 5.9 Comparison of the deep neural network with the sine function. Note that the DNN is computing a piecewise linear approximation

different lines defined on each of the intervals of length $\pi/2$ between 0 and 2π. The network is really just determining which of the four lines to use and then computing where on the line the input falls. A comparison between the neural network and the true function are shown in Fig. 5.9. This neural network is not the best possible neural network for approximating this function, but it does a reasonable job; in fact, this network was fit "by hand" for this example. If we trained a neural network using training data from the sine function, it is likely we would have obtained something very different.

5.3.2 Training Deep Neural Networks

Above we discussed the process of training a simple neural network using back propagation where the chain rule is used to get the derivative of the loss function with respect to a weight or bias. The process is the same for DNNs, but the details are significantly messier. This is due to the fact that in a DNN there are typically many weights and biases. Using our notation for a DNN the derivative of the squared-error loss function for a weight in layer ℓ in an L layer network is

$$\frac{\partial L}{\partial w_{ij}^{\ell \to \ell'}} = \delta_i^{(\ell')} z_j^{(\ell)}, \tag{5.29}$$

where $u_j^{(\ell')}$ is the value of node j in hidden layer ℓ' before applying the activation function $\sigma(u)$. In this case the value of $\delta_i^{(\ell')}$ is the sum of all of δ that $z_i^{\ell'}$ connects to

$$\delta_i^{(\ell')} = \begin{cases} \left(\displaystyle\sum_{k=1}^{K^{\ell'+1}} w_{kj}^{\ell' \to \ell'+1} \delta_k^{(\ell'+1)} \right) \dfrac{d\sigma}{du_i^{(\ell')}} & \ell' < L+1 \\[4mm] (\hat{y} - y) \dfrac{do}{du_o} & \ell' = L+1 \end{cases} . \tag{5.30}$$

Because the value for a derivative for weights in the initial layers depends on the deltas from the deeper layers, there can be problems with deltas growing too large or growing too small. If the deltas grow too large, then the updating procedure will be unstable, and, if the deltas are too small, then the network will not train. We will return to this topic later when we discuss recurrent neural networks in a future chapter.

There is also the issue of how to initialize the weights when training the model. Typically, the initial weights and biases are randomly drawn from a standard normal distribution (i.e., a normal distribution with mean zero and standard deviation one). Other approaches include the Xavier initialization scheme [1] where the weights are drawn from a normal distribution with mean zero and standard deviation equal to the square root of the number of input parameters. The initialization of weights is important because, as we can see in Eq. (5.30), the magnitude of the weights leads directly to the magnitude of $\delta_i^{(\ell')}$: too large initial weights make the network update unstable and too small initial weights make the network not update enough.

5.3.3   Neural Networks for Classification Problems

To this point we have only considered regression problems with feed-forward neural networks. Everything we have discussed can be applied to classification problems with only minor adjustments. The output layer for a K-class classification problem will be the softmax function (see Eq. (2.24)) that takes as input K different units from the final hidden layer and returns K probabilities, one for each class. For the loss function we can use the cross-entropy function (see Eq. (2.25)) or similar variants.

We only need to make changes to the loss function and have a final hidden layer. The other features of the network will be the same as regression: we have choices for the activation functions, the number of layers, and the number of units in the layers. The training is also the same, and we use gradient descent to decrease the loss function during the training.

5.4   Regularization and Dropout

Previously, we discussed regularization as a technique to reduce overfitting. The basic idea was to add a penalty term to the loss function so that more simple models are favored over more complicated models with a potential small loss of accuracy because now the loss function we minimize is not just a measure of error. Regularization is more important on neural networks than on linear models because the danger of overfitting is larger. We will discuss two important types of regularization here: weight decay and dropout.

In a neural network it is relatively easy to get thousands of weights in the network. With these many free parameters it is easy to make the weights overfit the training data. We want to make the weights in the network fit the training data, but also make the model as simple as possible. Therefore, we can add a term to the loss function when we fit the model. We can add an L_1 penalty that is analogous to lasso regularization by adding the sum of the absolute value of all the weights and biases in the network. Alternatively, we can add a penalty term that is an L_2 penalty, like in ridge regression, by adding the sum of the squares of the weights and biases to the loss function. Adding these penalties to the squared-error loss function for a single training point gives

$$L = (\hat{y} - y)^2 . + \lambda \left(\alpha \|\mathbf{w}\|_1 + (1 - \alpha) \|\mathbf{w}\|_2 \right), \qquad (5.31)$$

where $\lambda > 0$ is the strength of the regularization and $\alpha \in [0, 1]$ gives the balance between the L_1 and L_2 penalties (sometimes called weight decay). Both of these parameters need to be chosen by the user and can be chosen in a similar way as in regularized regression: λ is chosen to be the largest value that gives a model within one standard deviation of all the λ tested. Choosing α can be guided based on whether sparsity, that is having many zero weights and biases, is desired for a network. If sparsity is desired, α should be close to 1.

5.4.1   Dropout

Dropout is a regularization technique that is very popular for machine learning problems. The idea is to train random subsets of the neural network, rather than training the entire large network. Then we average these subsets when we wish to use the neural network to make a prediction. The random subset of the network is found by randomly setting to zero with probability p the value of a node (i.e., an input or hidden unit) and the connections to that node while training the model on a small subset of the training data. Then we repeat the process of randomly zeroing out some nodes and continuing to train. This makes it so that only a part of the network is trained on specific instances of the training data: some parts of the network will learn to predict some of the training data, while others are exposed to different training data. In this sense the network is a combination of many smaller networks each specialized to different data.

When we use a neural network trained using dropout, we do not zero out any of the nodes and multiply all the outgoing weights by the dropout probability p. This makes the network effectively an average of all the sub-networks used during training.

Dropout was first presented by Srivastava, Hinton, et al. [2] and they suggest that the probability of dropping out a unit should be about $p = 0.5$ for hidden units and closer to $p = 1$ for input units. Their paper shows that dropout works well with other regularization techniques (L_2 and L_1 decay, for example). They also find that dropout works with a variety of activation functions and network types. They also give different inspirations for dropout including the superiority of sexual reproduction to asexual reproduction to produce advanced species where different parents can specialize in different tasks, or the superiority of having many small conspiracies to one large conspiracy—10 small conspiracies of 5 people each are likely to have more success than a single large conspiracy of 50 people. A different biological explanation is that the human brain is using dropout all the time: our neurons misfire for all kinds of reasons; maybe our brains are a natural average of many "sub-brains." Whatever the motivation, dropout is a standard regularization technique for neural networks.

> **Regularization for Neural Networks**
> Given that a neural network can easily have thousands of parameters to fit for the weights and biases of the model, regularization is often necessary. Dropout and weight decay are two standard approaches. Dropout randomly sets to zero weights and biases during training to avoid overfitting. Weight decay is similar to regularized regression where we add to the loss function a penalty based on the magnitude of the weights. L_1 and L_2 weight decays behave like Lasso and ridge, respectively.

5.5   Case Study: The Strength of Concrete as a Function of Age and Ingredients

To demonstrate features of neural networks we will consider a data set consisting of a series of different types of concrete, based on ingredients and age, and the compressive strength of the concrete. This data comes from a paper by Yeh [3] and was retrieved from the UCI machine learning repository [4]. In the data there are eight independent variables: the age of the concrete in days, and seven variables corresponding to the mass of different components per unit volume of cement, water, blast furnace slag, fly ash, superplasticizer, fine aggregate, and coarse aggregate. The dependent variable for our study is the compressive strength of the concrete in MPa. There are 1030 cases in the data.

Figure 5.10 shows scatter plots of the compressive strength of the concrete versus each of the independent variables. From these plots we can see trends with respect to the age of the concrete (i.e., new concrete has reduced strength), water content (lower water content seems to indicate stronger concrete), and more cement tends to make stronger concrete. The standard deviation of the compressive strength in the data set is 16.698 MPa.

For our fitting we split the data randomly into a training and test set with 80% of the data being used in the training set. We also normalize the inputs by dividing each by the L_2 norm of each input.

5.5.1 Neural Networks with Raw Independent Variables

We begin by fitting a neural network with a single hidden layer of 8 inputs and an output layer with a single node for the compressive strength. There are 81 trainable parameters in this model: 72 weights and biases connecting the input layer to the hidden layer, and 9 weights and biases connecting the output layer to the hidden layer. We use the Adam optimization method for training [5], a type of stochastic gradient training algorithm, and train the model for 10^5 epochs with a batch size of 32. The activation function we use is the hyperbolic tangent function; this was chosen after some experimentation. We used Xavier initialization for the weights and biases and used an L_2 penalty of $\lambda = 0.0005$.

For the single hidden layer neural network we achieve a mean-absolute error, that is, the average absolute value of the difference between the true and predicted values, of 5.52 MPa on the test set. This model also has $R^2 = 0.800$ on the test set. From these numbers the model appears to be a useful predictor of the compressive strength given the 8 inputs. To further inspect the model we can look at the convergence of the mean-absolute error as a function of training epoch, as shown in Fig. 5.11a. In this figure we can see that the error is decreasing for both the training and test sets as the number of training epochs is increased. For this reason, we can be reasonably assured that the model is not overfit to the training data.

To attempt to improve on the results of a single hidden layer network, we fit a two hidden layer network. It would be possible to use a randomly initialized network and train it as we did with the single layer model, but in this case it makes sense to start from the one layer model. We create a new neural network; this one has two hidden layers; the first has 8 hidden units and the second has 2 hidden units. We initialize the weights and biases in the first layer to be the same as what were fit in the single layer network above. Then we train the other weights and biases on the same training set. The idea here is to find a model that is more complicated and can "correct" any deficiencies that we could not capture in a single layer model. The convergence of the error for this model is shown in Fig. 5.11b. Here again we see

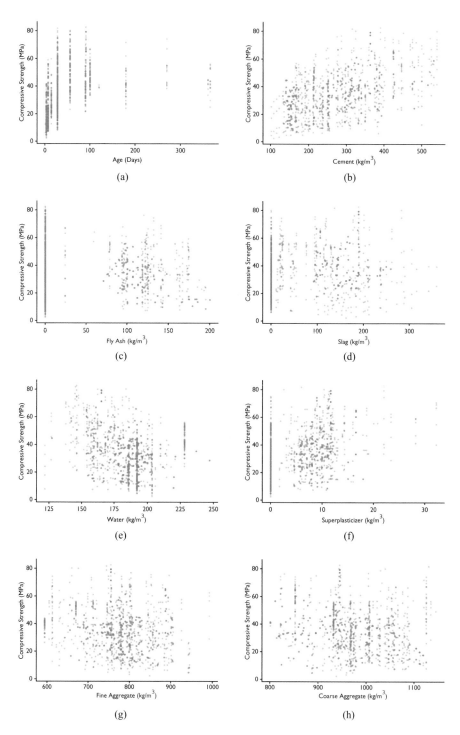

Fig. 5.10 Scatter plots showing the concrete strength versus the 8 different independent variables. (**a**) Age. (**b**) Cement. (**c**) Fly ash. (**d**) Slag. (**e**) Water. (**f**) Superplasticizer. (**g**) Fine aggregate. (**h**) Coarse aggregate

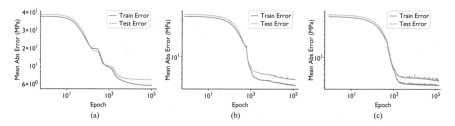

Fig. 5.11 Convergence of the mean-absolute error for the neural networks using the raw independent variables. (**a**) Single layer. (**b**) Two layers. (**c**) Three layers

that the error is decreasing in both the test and training sets as the number of epochs increases. It may be beneficial to continue training the model for more epics, but we stop here for this example. The mean-absolute error for the two-layer model is 4.96 MPa and $R^2 = 0.831$; both are an improvement over the single layer model.

If we continue and build a three hidden layer model by adding a new layer with two hidden units, and initialize this model with the weights and biases from the first two layers of the two hidden layer model, we get a model that performs worse on the test data. In this case we have a mean-absolute error of 5.05 MPa and $R^2 = 0.830$. Furthermore, we can see in Fig. 5.11c that the test error and training error have a much noisier convergence plot. This is likely due to the added complexity of the model: the three layer model has 99 parameters to fit.

This is an example of a more complicated model not necessarily performing better than the simpler model. Some of this is undoubtedly due to the fact that the optimization procedure does not find the absolute best model. It is minimizing the loss function, but it might not be finding the global minimum. If we retrain the model many times, we might be able to get a better model due to the random initialization of some of the weights and biases, but there is no guarantee.

5.5.2   Neural Networks with Additional Features

The original paper by Yeh defined a water to binder ratio as another independent variable and also used the logarithm of the age in days of the concrete. The water to binder ratio is a conventional parameter in studying concrete strength, as mentioned by Yeh. The water to binder ratio is defined as the amount of water divided by the sum of the cement and fly ash component. The water to binder ratio does seem to have a strong relationship to the compressive strength, as seen in the scatter plot of the compressive strength versus water to binder ratio in Fig. 5.12a. Using the water to binder ratio and the logarithm of the age in days as part of a one-layer model with all other network parameters the same (i.e., number of hidden units, activation function, initialization, and regularization) gives a mean-absolute error of 4.95 MPa and $R^2 = 0.831$. Initializing a two-layer model with the one-layer model, as we did

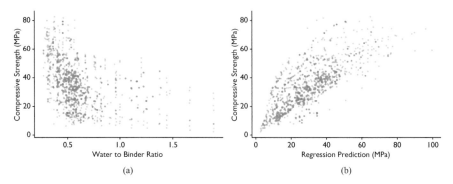

Fig. 5.12 Scatter plots showing the concrete strength versus the water to binder ratio and the results predicted by the regression model in [3]. (**a**) Water to binder ratio. (**b**) Regression model

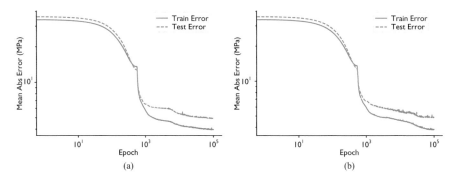

Fig. 5.13 Convergence of the mean-absolute error for the neural networks using the new independent variables of log days and water to binder ratio. (**a**) Single layer. (**b**) Two layers

above, gives a model with 4.84 MPa as the mean-absolute error and $R^2 = 0.834$. The convergence of these models is shown in Fig. 5.13. These results indicate that added new variables to the model, especially if they have physical meaning, can improve the model without adding much complexity.

The original paper containing the concrete data also fit a linear regression model to the compressive strength as

$$f_c = aX \left(c \ln t + d \right), \tag{5.32}$$

where f_c is the compressive strength and t is the age in days. Values for the regression coefficients $a, b, c,$ and d are given in the paper for 4 different randomized training sets. We use the average of these regression parameters to form a new independent variable to the model that is f_c: the predicted compressive strength from a linear regression model. The best value for R^2 on a test set for a linear model reported by Yeh is 0.779. In Fig. 5.12b we see that the regression prediction is directionally correct in that the regression model can qualitatively indicate the compressive strength, but the quantitative prediction can be inaccurate.

Adding the regression prediction as an independent variable to the model is a means to give an input that "close" to the correct answer. It is common to have approximate models that need to be corrected, and we want the neural network to learn this correction. We know that the linear regression model is missing some features of the data; we want the neural network to improve the linear regression model.

A one-layer model that includes all 11 independent variables (the original 8 inputs, the water to binder ratio, the log of age in days, and the linear regression prediction) gives a mean-absolute error of 5.13 MPa and $R^2 = 0.821$. This is an improvement over the one-layer model with only the raw data, but not an improvement of the model that includes the water to binder ratio and log age. Adding a second layer with two hidden units and initializing from the one-layer model give a mean-absolute error of 4.96 MPa and $R^2 = 0.836$. This two-layer model has the highest R^2 of any model tested, but the mean-absolute error is slightly worse than the two-layer model with the new independent variables.

What should we take away from this case study? First, we found that for this "small" data set of 1030 cases it is possible to predict the compressive strength of the concrete to within about 5 MPa without too much trouble. We can also improve on a linear regression model built with expert knowledge with simple neural networks. Our study was not comprehensive in that we could have tried more varied network architectures (different numbers of hidden units, etc.) and run more training epochs. We did learn, however, that deeper (that is more hidden layers) networks do not always outperform shallower networks.

Notes and Further Reading

There are many resources for further reading on neural networks, deep neural networks, and the issues with training them. The paper in *Nature* by LeCun, Bengio, and Hinton is a great read to see some of the application areas and basics of these networks [6]. Additionally, there are topics that we did not cover in this chapter that may be of interest to readers. One topic that we did not cover is batch normalization [7]. Batch normalization is an approach to speed up the training networks by making sure that the variations of the inputs within a batch are accounted for. Finally, there are whole books written about optimizers for training neural networks. Readers interested in learning more should see [8].

Problems

5.1 Consider a neural network with two hidden layers. By hand compute the derivative of the loss function with respect to a weight from each of the hidden layers and show that it is equivalent to Eq. (5.29).

5.2 Consider a neural network with one hidden layer. By hand compute the back propagation formula for a network regularized with L_1 and L_2 penalties as in Eq. (5.31).

5.3 Repeat problem 3.3 using neural networks of 1, 2, and 3 layers. How accurate can you get your models? Plot your results as a function of z at time $\tau = 10$.

5.4 Repeat problem 5.3 by building models that include a) dropout, b) L_2 regularization, and c) L_1 regularization.

5.5 Build neural networks of 1, 2, and 3 layers to approximate the Rosenbrock function

$$f(x, y) = (1 - x)^2 + 100(y - x^2)^2,$$

by sampling 100 points randomly in the interval $[-1, 1]$ to create the training set. Try several different activation functions and regularizations. Use the models to find the location of the minimum of the function and compare your result with the true value of $x = -1$ and $y = 1$.

References

1. Xavier Glorot and Yoshua Bengio. Understanding the difficulty of training deep feedforward neural networks. In *Proceedings of the Thirteenth International Conference on Artificial Intelligence and Statistics*, pages 249–256, 2010.
2. Nitish Srivastava, Geoffrey Hinton, Alex Krizhevsky, Ilya Sutskever, and Ruslan Salakhutdinov. Dropout: a simple way to prevent neural networks from overfitting. *The Journal of Machine Learning Research*, 15(1):1929–1958, 2014.
3. I-C Yeh. Modeling of strength of high-performance concrete using artificial neural networks. *Cement and Concrete Research*, 28(12):1797–1808, 1998.
4. Arthur Asuncion and David Newman. UCI machine learning repository, 2007.
5. Diederik P Kingma and Jimmy Ba. Adam: A method for stochastic optimization. *arXiv preprint arXiv:1412.6980*, 2014.
6. Yann LeCun, Yoshua Bengio, and Geoffrey Hinton. Deep learning. *Nature*, 521(7553):436–444, 2015.
7. Sergey Ioffe and Christian Szegedy. Batch normalization: Accelerating deep network training by reducing internal covariate shift. *arXiv preprint arXiv:1502.03167*, 2015.
8. Sebastian Ruder. An overview of gradient descent optimization algorithms. *arXiv preprint arXiv:1609.04747*, 2016.

Chapter 6
Convolutional Neural Networks for Scientific Images and Other Large Data Sets

Yea, the very sight of them was so terrible, so fearful, and so dreadful that all my hair stood on end, and I could believe nothing but that they were all bereft of reason.

—Hans Jakob Christoffel von Grimmelshausen Simplicius Simplicissimus, *translated by Alfred Thomas Scrope Goodrick*

Abstract For problems where there is a large number of inputs, such as an image where each pixel can be considered an input, a feed-forward neural network would have a truly huge number of weight and bias parameters to fit during training. For such problems rather than considering each input to be independent, we take advantage of the fact that the input has structure, even if we do not know what that structure is, by using convolutions. In a convolution we apply a function of particular form repeatedly over the inputs to create the hidden layer variables. In image data, this allows the network to highlight important features in an image. We begin the discussion of convolutions by defining and giving examples of 1-D convolutions for functions and for vectors, before extending to multiple dimensions. We then talk about a special type of operation applied in machine learning called pooling, before showing how convolutions work in neural networks. We apply convolutional neural networks to the Fashion MNIST data set and give a case study of using these networks, along with a concept called transfer learning, to find volcanoes on Venus.

Keywords Convolutional neural networks (CNN) · Convolutions · Discrete convolutions · Pooling · Transfer learning

6.1 Convolutions

The neural networks that we discussed in the previous chapter had many possible unknown parameters (i.e., the weights and biases) to train. The number of parameters increases as the number of inputs is increased and as the network gets deeper by adding more hidden layers. In this chapter we investigate how we can reduce the

© Springer Nature Switzerland AG 2021
R. G. McClarren, *Machine Learning for Engineers*,
https://doi.org/10.1007/978-3-030-70388-2_6

number of parameters to fit when we have a large number of inputs to the model, as, for example, if we have an image as the input to the model. This is done by making sets of weights have the same values and applying those weights repeatedly in a pattern.

We begin with the definition of a convolution of two functions. Consider two functions f and g; the convolution $f * g$ is defined as

$$(f * g)(t) = \int_{-\infty}^{\infty} f(\tau)g(t - \tau)\, d\tau = \int_{-\infty}^{\infty} f(t - \tau)g(\tau)\, d\tau. \tag{6.1}$$

Note that the definition is commutative, $f * g = g * f$. The convolution is linear in that if k is a scalar $(kf) * g = k(f * g)$ and $(f_1 * g) + (f_2 * g) = (f_1 + f_2) * g$; also, the convolution is shift invariant: this means that if one shifts the argument of one of the functions by a scalar k, the convolution does not change. If $\tilde{f}(t) = f(t + k)$, then $\tilde{f} * g = f * g$.

If we consider the special case of the function g having finite support, i.e., $g(t) = 0$ for $|t| > a$, we can simplify the integral:

$$(f * g)(t) = \int_{-\infty}^{\infty} f(t - \tau)g(\tau)\, d\tau \tag{6.2}$$

$$= \int_{-a}^{a} f(t - \tau)g(\tau)\, d\tau.$$

This shows that the convolution evaluated at t is the integral of f around the point t times the function g.

Oftentimes the convolution will produce a smoothed version of the function $f(t)$ if $g(t)$ is a function that has finite support. Some functions that perform a smoothing are the rectangular function,

$$g_{\text{rect}}(t) = \frac{1}{2a}(H(t + a) - H(t - a)), \tag{6.3}$$

where $H(t)$ is the Heaviside unit step function, and the triangle function

$$g_{\text{tri}}(t) = \frac{1}{a^2}(a - |t|)_+, \tag{6.4}$$

where again the plus subscript denotes that the function returns zero if its argument is negative. If $g(t)$ is the rectangular function, the convolution becomes a simple integral:

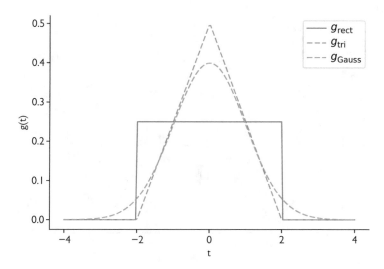

Fig. 6.1 Plot of g_{rect}, g_{tri}, and g_{Gauss} with $a = 4$ and $\sigma = 2$

$$(f * g_{rect})(t) = \frac{1}{2a} \int\limits_{-a}^{a} f(t - \tau) \, d\tau. \tag{6.5}$$

This convolution replaces $f(t)$ with the average of $f(t)$ over a finite range. As integrals/averages are almost always smoother than the original function, $f * g_{rect}$ is a smoothed version of $f(t)$. Similarly, when g is the triangular function, we get an average of f that is weighted so that points near t are more strongly weighted.

We could define many possible functions g that act as a weighting function in averaging f, and thereby smoothing it. An extreme example is the Gaussian function that exponentially approaches zero:

$$g_{Gauss}(t) = \frac{1}{\sqrt{2\pi\sigma^2}} e^{-t^2/(2\sigma^2)}, \tag{6.6}$$

where $\sigma > 0$ is a parameter. The integral of the Gaussian over the entire real line is 1 and it has a maximum at $g_{Gauss}(0)$, so it will produce a weighted average of f that is centered at t.

In Fig. 6.1 we plot the values of the different example g functions defined above. We see that with $a = 2\sigma$ the Gaussian and triangular functions have roughly the same shape, with the Gaussian function more smoothly transitioning to zero and not reaching as high of a value. We will see that convolutions with these two functions behave in a similar manner.

Because convolution smooths functions, they can be useful for treating noisy data or functions before differentiating them. For example, if one wants to take a derivative of a signal, the differentiation operation will amplify the noise. If we

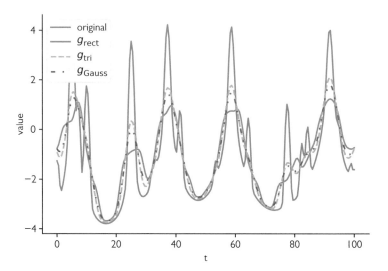

Fig. 6.2 Demonstration of the smoothing of a signal $f(t)$ and using convolutions with g_{rect}, g_{tri}, and g_{Gauss} with $a = 4$ and $\sigma = 2$

perform a smoothing convolution first, we can improve the derivative estimate. This behavior is shown in Fig. 6.2 where a noisy signal is convolved with the three g functions defined above using $a = 4$ and $\sigma = 2$ to get a much smoother function. Here we see that the rectangular function smooths the function more drastically, making the peaks much lower. All three functions remove the narrow peaks that appear next to the "main" peak, but the triangular and Gaussian functions result in a convolution that better preserves the shape of the large peaks.

The derivative of a convolution is related to the convolution of a derivative. To see this we mechanically apply the derivative to a convolution:

$$\frac{d}{dt}(f * g)(t) = \frac{d}{dt} \int_{-\infty}^{\infty} f(\tau)g(t - \tau)\,d\tau, \tag{6.7}$$

$$= \int_{-\infty}^{\infty} f(\tau)\frac{d}{dt}g(t - \tau)\,d\tau,$$

$$= \left(f * \frac{d}{dt}g\right)(t) = \left(\frac{d}{dt}f * g\right)(t).$$

Therefore, we can get the derivative of a convolution by performing a convolution with the derivative of the smoothing function. If we use the rectangle function, we can see that the derivative of the convolution is the negative of the finite difference derivative:

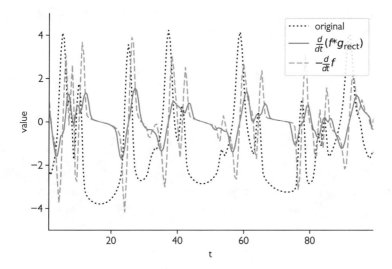

Fig. 6.3 Demonstration of the approximate derivative calculated by taking the derivative of a convolution with the rectangular function, g_{rect}, and $f(t)$ using $a = 2$

$$\frac{d}{dt}(f * g_{rect})(t) = \frac{1}{2a} \int_{-\infty}^{\infty} f(t - \tau)(\delta(t + a) - \delta(t - a)) \, d\tau, \qquad (6.8)$$

$$= \frac{f(t - a) - f(t + a)}{2a}.$$

When we use other weighting functions, we get smoother versions of the derivative, but all of them require convolving with the derivative of the weighting function.

We demonstrate that $\frac{d}{dt}(f * g_{rect})(t)$ gives the negative of the finite difference estimate of the derivative in Fig. 6.3. Here we plot the derivative of the convolution as well as the true value of -1 times the derivative of $f(t)$. In the derivative of the signal we see a large negative value followed by a large positive value when the original signal goes through a peak. The result from $\frac{d}{dt}(f * g_{rect})(t)$ smooths the derivative noticeably. This is a result of the signal being smoothed by convolving with g_{rect} before taking the derivative. This is important if we believe that there is noise in the signal, but we need to compute the derivative.

Before continuing on, we point out that convolutions can be extended to d dimensions in a straightforward manner. If f and g are functions of $x_1, \ldots, x_d$, then the convolution is

$$(f * g)(x_1, \ldots, x_d) = \int_{-\infty}^{\infty} d\hat{x}_1 \ldots \int_{-\infty}^{\infty} d\hat{x}_d \, f(x_1, \ldots, x_d) g(x_1 - \hat{x}_1, \ldots, x_d - \hat{x}_d).$$

$$(6.9)$$

The commutativity and linearity of the convolution also hold in multiple dimensions. The special case of $d = 2$ is given by

$$(f * g)(x_1, x_2) = \int_{-\infty}^{\infty} d\hat{x}_1 \int_{-\infty}^{\infty} d\hat{x}_2 \, f(x_1, x_2) g(x_1 - \hat{x}_1, x_2 - \hat{x}_2). \qquad (6.10)$$

Convolutions
A convolution involves the integral of the product of functions. This operation can smooth a function under many circumstances and can provide a way to take a derivative of a noisy function without amplifying the noise.

6.1.1 Discrete Convolutions

In many cases we do not have a continuous function to work with and we must deal with the function at a finite number of points. In this case we consider a convolution of two vectors. If $\mathbf{f}$ is a vector of length N, we write the nth component of the discrete convolution, with subscripts denoting vector elements with zero-based indexing,[1] as

$$(\mathbf{f} * \mathbf{g})_n = \sum_{m=-\infty}^{\infty} f_m g_{n-m} = \sum_{m=-\infty}^{\infty} f_{n-m} g_m \qquad n = 0, \ldots, N - 1, \qquad (6.11)$$

with the convention that all indices out of bounds for $\mathbf{f}$, e.g., negative values or values greater than $N - 1$, are zero. This is called zero padding: we add enough zeros to the ends of $\mathbf{f}$ so that the result of the convolution has the same size as the original vector. We can define the vector $\mathbf{g}$ to have negative indices, however, and this will be worthwhile, as an example shows. The vector $\mathbf{g}$ that is used in the convolution is often called a "kernel." The size of the kernel is number of entries in g.

The discrete convolution can be used to define smoothing in a similar way to the continuous case. For example, if we set

$$g_{\text{avg},m} = \begin{cases} \frac{1}{3} & m = -1, 0, \text{ or } 1 \\ 0 & \text{otherwise} \end{cases}, \qquad (6.12)$$

---

[1] We need to use zero-based indexing for our vectors for the formulas in this section to make sense without adding several non-intuitive constants to our indices.

the discrete convolution will be the average of three points:

$$(\mathbf{f} * \mathbf{g}_{avg})_n = \frac{1}{3}(f_{n+1} + f_n + f_{n-1}). \tag{6.13}$$

In this case the kernel $\mathbf{g}_{avg}$ has a size of 3.

Note that the way we have defined the discrete convolution, near the end of the vector $\mathbf{f}$, the formulas will be slightly different. For instance, the convolution that gives the average of $\mathbf{f}$ will behave differently near the edge:

$$(\mathbf{f} * \mathbf{g}_{avg})_0 = \frac{1}{3}(f_1 + f_0), \quad (\mathbf{f} * \mathbf{g}_{avg})_{N-1} = \frac{1}{3}(f_{N-1} + f_{N-2}). \tag{6.14}$$

If the vector is long, i.e., N is large, these edge effects will have a small impact on the result from the discrete convolution. We could also make other changes, such as making the vector periodic by setting $f_{-1} = f_{N-1}$ and $f_N = f_0$, or by changing $\mathbf{g}$ at the boundary. In general, we will be able to ignore these edge effects in our use of convolutions.

We can also make the convolution give us a finite difference approximation of a second derivative. If we define

$$g_{D2,m} = \begin{cases} \frac{1}{h^2} & m = -1, \text{ or } 1 \\ -\frac{2}{h^2} & m = 0 \\ 0 & \text{otherwise} \end{cases}, \tag{6.15}$$

the discrete convolution will be the finite difference approximation to the second derivative of f:

$$(\mathbf{f} * \mathbf{g}_{D2})_n = \frac{1}{h^2}(f_{n+1} - 2f_n + f_{n-1}). \tag{6.16}$$

6.1.2 Two-Dimensional Convolutions

In machine learning two-dimensional convolutions are common, especially when dealing with image data. In this case we consider $\mathbf{F}$ and $\mathbf{G}$ to be matrices. In this case we write the convolution as the sum over the product of matrix elements:

$$
\begin{aligned}
(\mathbf{F} * \mathbf{G})_{m,n} &= \sum_{i=-\infty}^{\infty} \sum_{j=-\infty}^{\infty} F_{i,j} G_{m-i,n-j} \\
&= \sum_{i=-\infty}^{\infty} \sum_{j=-\infty}^{\infty} F_{m-i,n-j} G_{i,j} \quad m, n = 0, \dots, N-1.
\end{aligned} \tag{6.17}
$$

In this case we can think of the matrix $\mathbf{G}$ as giving a stencil to average $\mathbf{F}$ over. For instance if $\mathbf{G}$ is

$$\mathbf{G}_{\text{mean}} = \frac{1}{9} \begin{pmatrix} 1 & 1 & 1 \\ 1 & 1 & 1 \\ 1 & 1 & 1 \end{pmatrix}, \tag{6.18}$$

the convolution $(\mathbf{F} * \mathbf{G})_{m,n}$ gives the average of the nine values near $F_{m,n}$:

$$(\mathbf{F} * \mathbf{G}_{\text{mean}})_{m,n} = \frac{1}{9} \sum_{i=-1}^{1} \sum_{j=-1}^{1} F_{m-i,n-j}. \tag{6.19}$$

In practice this kernel $\mathbf{G}_{\text{mean}}$ is called the mean filter of size 3×3. We could define mean filters of different sizes that would average $\mathbf{F}$ over a larger area.

Also, we can take the discrete Laplacian $\Delta^2 = \frac{\partial^2}{\partial x^2} + \frac{\partial^2}{\partial y^2}$ of the signal with the proper definition of $\mathbf{G}$:

$$\mathbf{G}_{\text{Lap}} = \frac{1}{h^2} \begin{pmatrix} 0 & 1 & 0 \\ 1 & -4 & 1 \\ 0 & 1 & 0 \end{pmatrix}. \tag{6.20}$$

This makes the convolution compute the finite difference approximation of the Laplacian

$$(\mathbf{F} * \mathbf{G}_{\text{Lap}})_{m,n} = \frac{1}{h^2} \left[(F_{m-1,n} - 2F_{m,n} + F_{m+1,n}) + (F_{m,n-1} - 2F_{m,n} + F_{m,n+1}) \right]. \tag{6.21}$$

When considering discrete convolutions we need to consider what happens at the edge of the data. Above we mentioned that we add zeros around the edges of the data so that the convolution result has the same size as the original data. Alternatively, one cannot add any zeros so that the data size shrinks because the convolution cannot be applied at the edges. To demonstrate this we consider the data

$$\mathbf{F} = \begin{pmatrix} 1 & 2 & 3 \\ 4 & 5 & 6 \\ 7 & 8 & 9 \end{pmatrix},$$

and we want to apply the mean filter to this data given by Eq. (6.18). If we do not use padding, then there is only one entry in $\mathbf{F}$ that we can apply the convolution to, the value 5 in the middle:

$$\mathbf{F} * \mathbf{G}_{\text{mean}} = \begin{pmatrix} 5 \end{pmatrix} \qquad \text{No padding.}$$

If we pad the data with a row or column on each side, we get a result that has the same size as $\mathbf{F}$:

$$\mathbf{F} * \mathbf{G}_{\text{mean}} = \begin{pmatrix} 0\,0\,0\,0\,0 \\ 0\,1\,2\,3\,0 \\ 0\,4\,5\,6\,0 \\ 0\,7\,8\,9\,0 \\ 0\,0\,0\,0\,0 \end{pmatrix} * \mathbf{G}_{\text{mean}} = \begin{pmatrix} \frac{4}{3} & \frac{7}{3} & \frac{16}{9} \\ \frac{25}{9} & 5 & \frac{11}{3} \\ \frac{26}{9} & \frac{13}{3} & \frac{28}{9} \end{pmatrix} \qquad \text{With padding.}$$

The number of zeros that need to be added in the padding depends on the size of the convolution matrix (in this case $\mathbf{G}$). If the matrix $\mathbf{F}$ is large, the effects at the edge may be negligible and the issue of padding choice may not matter.

Figure 6.4 demonstrates the application of convolutions in 2-D. This convolution has a 3×3 matrix for $\mathbf{G}$. The top portion of the figure shows $\mathbf{F}$, $\mathbf{G}$, and $\mathbf{F} * \mathbf{G}$. The middle panel shows how 9 entries in $\mathbf{F}$ are combined to get an interior entry, and the bottom row panel demonstrates the necessity of padding to keep the size of the result identical to the original matrix.

For a further example of 2-D convolutions we will consider an image and apply different convolutions to it. In Fig. 6.5a we show a grayscale image normalized so that each pixel takes on a value between 0 and 1. When we apply the 2-D Laplacian from Eq. (6.21) to the image, the result is Fig. 6.5b. The resulting image seems to highlight the edges of the original image. We see the outline of the figures and objects in the image; this demonstrates that the Laplacian convolution can find where there are sharp changes in the data.

The Laplacian convolution is sensitive to noise in the data however. In Fig. 6.5c the original image has had noise added to it in the form of a Gaussian with mean 0 and standard deviation 0.025. The noise is hardly perceptible to the eye, but when the Laplacian convolution is applied to the image, as shown in Fig. 6.5d, we see that the result has amplified noise and that the edges of the image are not perceptible. We can remedy the situation by first applying a mean filter. In Fig. 6.5e we apply a mean convolutional of size 9×9 to the noisy image. The resulting image is said to have a mean filter applied, and the result does have noticeable smoothing. However, when the mean-filtered image has the Laplacian convolution applied to it, we can again see the edges of the image in Fig. 6.5f. The result does have the artifact that the edges have spread out, i.e., they are thicker, compared with Fig. 6.5b; nevertheless, the result is superior to the Laplacian applied to the noisy image.

6.1.3 Multi-Channel 2-D Convolutions

Many of the ideas behind convolutional neural networks come from image processing. Many images have multiple channels with 3 being the most common: a

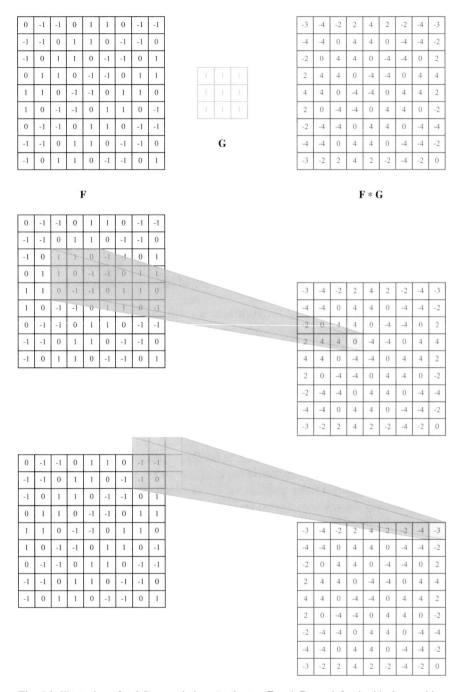

Fig. 6.4 Illustration of a 2-D convolution. At the top **F** and **G** are defined with the resulting convolution shown at the top right. The middle of the figure shows how **G** is applied to get an entry in **F** ∗ **G**, and the bottom portion demonstrates how the convolution needs padded values to preserve the size of **F**

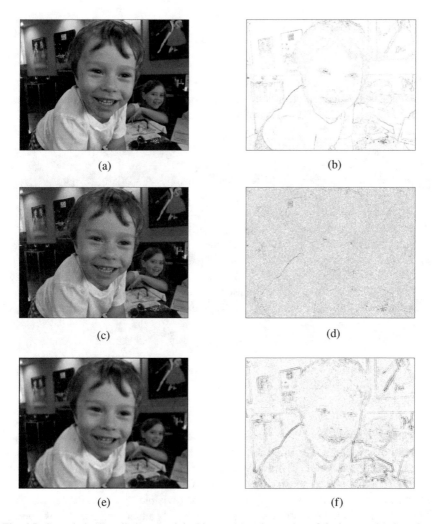

Fig. 6.5 Convolutions applied to an original image (**a**) and a version of the image with Gaussian noise. (**a**) Original image. (**b**) 2-D Laplacian convolution of original. (**c**) Original image with noise. (**d**) 2-D Laplacian convolution of noisy image. (**e**) Mean convolution on noisy image. (**f**) 2-D Laplacian convolution of mean-filtered noisy image. Photo credit: the author

channel for the red, green, and blue levels in the image. This means that an image is actually 3 matrices. When applying a convolution to this type of input, a different convolution is applied to each matrix and the results are summed up. To define this convolution we consider C channels stored in a tensor of size $M \times N \times C$ (i.e., C is the number of matrices of size $M \times N$). Call this tensor $\underline{\mathbf{F}}$ and we wish to convolve it with another tensor $\underline{\mathbf{G}}$. The convolution will be the sum of C 2-D convolutions

$$\underline{\mathbf{F}} * \underline{\mathbf{G}} = \sum_{c=1}^{C} \mathbf{F}_c * \mathbf{G}_c, \tag{6.22}$$

where the subscript c denotes a channel in the tensor.

Multi-channel convolutions are an important feature in convolutional neural networks when combined with pooling, a topic we discuss next.

Discrete Convolutions
In machine learning we typically deal with inputs and outputs that are of a finite number of dimensions. In this case convolutions turn into operations on vectors or matrices depending on whether the data layout is 1-D or 2-D. Whereas continuous convolutions involve multiplication and integration, discrete convolutions look like elementwise multiplication and summation. When applying discrete convolutions we can smooth a signal or take its derivative as before. Additionally, we usually have to pad the data if we want to get an output from the convolution that is the same size as the input. One of the most important features of a discrete convolution is that we can define the kernel $\mathbf{G}$ in 2-D of size $K \times K$ using only K^2 values.

6.2   Pooling

In machine learning "pooling" refers to the process of applying a function to a signal without overlapping. It is often used to take the average over part of the signal or find the maximum value. The average pooling operation applied to a 1-D vector of size N requires one to specify the "stride" or the number of elements to take the average over. Typically one wants the N to be divisible by the stride length, S. If this is not the case, the vector can have zero elements added to the end of the vector to make the vector length a multiple of S. The average pooling operation of stride S is given by[2]

$$\text{AvgPool}(\mathbf{x})_k = \frac{1}{S} \sum_{i=(k-1)S+1}^{kS} x_i, \qquad k = 1, \ldots, N/S. \tag{6.23}$$

The length of the resulting vector is N/S, i.e., the pooling reduces the length of the vector by a factor of S. As an example consider a vector of length 9 given by

---

[2]We have gone back to 1-based indexing of vectors in this section.

$$\mathbf{x} = (1, 1, 1, 1, 10, 1, 1, 1, 1);$$

applying average pooling with a stride of 3 results in a length 3 vector:

$$\text{AvgPool}(\mathbf{x}) = (1, 4, 1).$$

In this example, the input vector had a large value in the middle of the vector and resulting from the average pooling operation did as well. Notice that if we shift the location of the large value by one index either way (to make the large value in the fourth or sixth position in vector), the value for the average pooling operation does not change. That is, if we shift to the left or right (denoted by a $\mp$ subscript),

$$\mathbf{x}_- = (1, 1, 1, 10, 1, 1, 1, 1, 1) \quad \text{or} \quad \mathbf{x}_+ = (1, 1, 1, 1, 1, 10, 1, 1, 1),$$

the average pooling values are identical to the original vector

$$\text{AvgPool}(\mathbf{x}) = \text{AvgPool}(\mathbf{x}_-) = \text{AvgPool}(\mathbf{x}_+) = (1, 4, 1).$$

However, if we shifted the position of the large value in the vector too far, the average pooling result would change.

This ability of average pooling to be somewhat insensitive to the exact position of features in the input vector is one of the reasons they are valuable in machine learning. There are many cases where shifting a vector by a small amount should not affect the behavior of a machine learning model, and average pooling helps add this insensitivity. The insensitivity to shifts in the input is a function of the size of the stride, S. If S is large, then one can shift the input by a lot and not change the result much with the side effect of reducing the size of the signal more. When S is small, the pooling result will be more sensitive to shifts in the data.

The other common type of pooling called "max" pooling involves finding the maximum value in the stride. It behaves in much the same way as the average filter. The max pooling operation of stride S is

$$\text{MaxPool}(\mathbf{x})_k = \max(x_{(k-1)S+1}, \dots, x_{kS}) \qquad k = 1, \dots, N/S. \qquad (6.24)$$

The result from our example above gives slightly different results using the max pooling than with average pooling, but the character of the result is similar: with

$$\mathbf{x} = (1, 1, 1, 1, 10, 1, 1, 1, 1),$$

the result from max pooling is

$$\text{MaxPool}(\mathbf{x}) = (1, 10, 1).$$

Notice that max pooling preserved the largest value of the input vector, whereas average pooling reduced it through the averaging process. In this example shifting by one element does not change the value, as with average pooling:

$$\mathrm{MaxPool}(\mathbf{x}) = \mathrm{MaxPool}(\mathbf{x}_-) = \mathrm{MaxPool}(\mathbf{x}_+) = (1, 10, 1).$$

There is a special name for pooling operations when the stride S is equal to the vector length N. These operations are called global pooling operations and they reduce the input to a single value. Clearly, global pooling is not sensitive to shifting the inputs.

6.2.1   2-D Pooling

The application of pooling to two-dimensional data, i.e., a matrix or image, is a common application of pooling. In this case the pooling stride defines a double sum over the rows and columns of the data. For data of size $M \times N$, a stride S pool will result in a matrix of size $M/S \times N/S$. The average pooling operation with stride S in 2-D is defined as

$$\mathrm{AvgPool}(\mathbf{X})_{k\ell} = \frac{1}{S^2} \sum_{i=(k-1)S+1}^{kS} \sum_{j=(\ell-1)S+1}^{\ell S} X_{ij}, \quad k=1,\ldots,M/S, \ \ \ell=1,\ldots,N/S.$$
$$(6.25)$$

Similarly, the max pooling operation is

$$\mathrm{MaxPool}(\mathbf{X})_{k\ell} = \max(X_{(k-1)S+1,(\ell-1)S+1}, \ldots, X_{kS,(\ell-1)S+1}, \ldots X_{kS,\ell S}),$$
$$k=1,\ldots,M/S, \quad \ell=1,\ldots,N/S. \quad (6.26)$$

Before seeing an example of pooling, we point out that there is an operation known as strided convolution that combines pooling and convolution into one operation. In this case the convolution is not applied to each data point. Rather, the convolution is applied and then is moved in a way that skips a given number of points in the horizontal and vertical directions before applying the convolution again. The number of points skipped in a direction is called the stride, just as in a pooling layer, and the resulting data will have its size reduced by a factor of stride in each direction. Such a strided convolution does not apply a pooling operation, but the net result is reduced data size in a similar manner to pooling.

We can apply pooling to the image from Fig. 6.5 to demonstrate the effects of this operation. In Fig. 6.6 we apply max pooling of stride 9 to the Laplacian-convolved images shown in Fig. 6.5b, f. Though the images before applying the pooling have noticeable differences, e.g., there is much more detail in Fig. 6.6a, after applying max pooling, the two images have similar features. For instance, we could easily find the location of eyes in each of the images. Note that the number of pixels in the image is smaller by a factor of 9 in each direction after applying the max pooling operation.

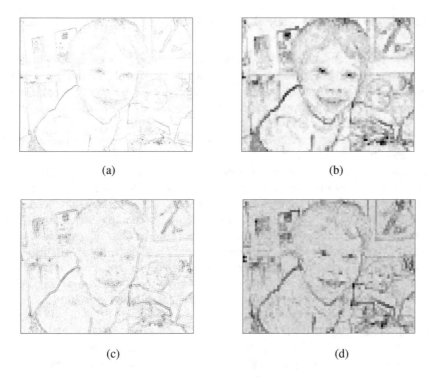

(a)   (b)

(c)   (d)

Fig. 6.6 Max pooling of stride 9 applied to the Laplacian-convolved images from Fig. 6.5. (**a**) 2-D Laplacian convolution of original. (**b**) Max pooling of (**a**). (**c**) 2-D Laplacian convolution of mean-filtered noisy image. (**d**) Max pooling of (**c**)

The characteristics of 2-D pooling is similar to those of 1-D pooling. As with 1-D pooling operations, 2-D global pooling returns a single number. It is possible to have the stride be different in the two different directions and sometimes this is necessary. We will now look at how pooling and convolution combine to pull out features from data.

Pooling
Pooling is an operation that can be applied to vectors or matrices to find features in the data. The pooling operation decreases the size of the signal and has a result that is somewhat insensitive to the exact location of a feature in the signal. This insensitivity will be used by machine learning models. Common types of pooling include average pooling that returns the average of the input over a particular patch of the input and the max pooling operation that returns the maximum value over a patch of the input.

6.3   Convolutional Neural Networks

If we combine convolutions and pooling, we can construct a neural network that can have a large input space but require a modest number of trainable parameters. Our discussion will focus on 2-D inputs because these are the most common uses of convolutional neural networks. Other data shapes are possible and our presentation is readily extendable to them.

We consider a single channel image of size $M \times N$ that we wish to use as input to a classification problem (e.g., we want to know if one of several objects is in the image). The standard procedure is to combine several layers of convolutions and pooling to *increase* the number of channels and *decrease* the size of the image before passing the smaller image over many channels to one or several fully connected feed-forward layers. Each convolution or pooling operation is passed through an activation function; this adds nonlinearity to the network. The fully connected feed-forward layers are sometimes called dense layers. These dense layers will also have a bias included in them, and the convolutional layers can have a bias as well. The dense layers will compute an output from the image. This output could be a number for a regression problem or a softmax layer for a classification problem.

A schematic of a convolutional neural network is shown in Fig. 6.7. In the figure, the input is an image that has three convolutions applied to it. The result of the three convolutions is three versions of the image that have been processed through the convolution. Then pooling layers shrink those three channels down to smaller sizes. The third layer is another convolution layer with five convolutions. Each convolution is applied to all three of the results of the pooling layer. Therefore, if the convolution is a 5×5 convolution, it would have $3 \times (5 \times 5) = 75$ unknowns in the convolution because each layer input to the convolution can have its own values. This second convolution layer has five channels that are then shrunk in a final pooling layer before being passed to the dense layer(s) and the output.

In the figure we can see that the image width is shrunk over subsequent layers but the number of channels is increased. We will see this architecture repeated in the next case study.

> **Convolutional Neural Networks**
> Using convolutions and pooling, convolutional neural networks (CNNs) can handle large amounts of input data with a relatively small number of parameters to train. When applied to matrices or images, the output of each convolutional or pooling layer is another matrix/image. After applying several of these layers to extract features from the data, we can pass a reduced-sized matrix to a series of fully connected feed-forward layers to get an output.

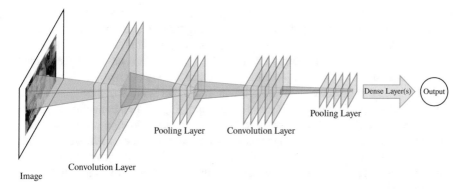

Fig. 6.7 Schematic of a typical convolutional neural network. The number of planes at each layer indicates the number of channels. The dense layers before the output nodes are not shown. The number of planes in a layer indicates the number of channels, and the plane area represents the size of the data

6.4 Case Study: Classification with Fashion MNIST

We have already seen the fashion MNIST data set in Chap. 4. It is a set of grayscale images of size 28×28 each containing one of ten types of clothing articles (e.g., shirt, sneaker, dress, boot, etc.). Here we use a convolutional neural network to try to classify these images. There are many examples of CNNs applied to this problem available and the idea for the network structure we use comes from [1].

Our network begins with a convolutional layer applied to the original 28×28 image. This convolution layer has 64 channels and each convolution is a 2×2 convolution. This first convolution layer then has $2 \times 2 \times 64 + 64 = 320$ unknown parameters. The additional 64 are for a bias in each channel. This layer is then followed by a max pooling layer with stride 2. After the max pooling layer there are 64 channels each with a 14×14 image. The max pooling layer is followed by a dropout layer that randomly sets some of the values in the 64 channels to zero.

Next, there is a convolutional layer of size 2×2 applied to create 32 channels. Each of 32 convolutions is 2×2 for each of the 64 input channels so that the total number of parameters is $(2 \times 2) \times 64 \times 32 + 32 = 8\,224$. The result from this second convolutional layer is 32 channels each with a size 14×14. This passes to a max pooling layer with stride 2 and again to a dropout layer, resulting in 32 channels each with size 7×7. This then passes to a fully connected layer with 256 neurons and the ReLU activation function. In the fully connected layer there are $256 \times 32 \times 7 \times 7 + 256 = 401\,664$ parameters. Dropout is applied before a final fully connected layer that with 10 neurons to compute the softmax output for the 10 different classes of images. This final layer has $256 \times 10 + 10 = 2\,570$ weights to fit.

This network overall has 412 778 parameters to fit. We fit it with the Adam optimizer and train it with a batch size of 64 for 10 epochs using 60 000 training cases and 10 000 test cases. After training the model has a training accuracy of 87.88% and a test accuracy of 89.98%, indicating that we can expect the model to correctly determine what type of item is in the picture about 90% of the time.

Our real interest in this example is how it illustrates the behavior of the trained convolution layers. In Fig. 6.8 we show the values of the intermediate channels for two images. At the top we show the original image that is input to the CNN. Below that we show the 64 channels that result from applying the first convolutional layer (the images are arranged in a 4×12 grid). Then the max pooling layer is applied, and the result is a shrinking of each channel by a factor of 2 in each dimension. Then the results of the 32 channel convolution are shown below that followed by another max pooling layer. The result of this layer is then fed to the fully connected layers and then the output.

Looking at Fig. 6.8 we see what the convolutions are doing: they are detecting the edges of the items. For instance, in the coat we see that the shoulders of the coat are highlighted and then in the final layer before the fully connected layer, the locations of the dark patches indicate the overall shape of the item. The sandal at the bottom of the figure has different areas highlighted in this final layer: it is these patterns the model uses to distinguish items.

6.5 Case Study: Finding Volcanoes on Venus with Pre-fit Models

One of the challenges of convolutional neural networks is that they can take a long time to train and can be difficult to properly design (i.e., choose the convolution sizes, etc.) for a given task. For this reason it is often productive to begin with a network that has been pre-trained on some representative task and augment it to solve the problem at hand. This will often take the form of loading a pre-trained model and adding a number of fully connected layers to the end of the network and training *just those added layers* on the task at hand, freezing the weights in the pre-trained network.

In this case study we will do just that using the EfficientNet network developed by Google [2]. The pre-trained version of EfficientNet that we use has been trained to solve the "ImageNet" classification problem where an image is classified as one of 1000 classes [3]. The version of EfficientNet that we use takes in an image of size 240×240 with 3 color channels (i.e., the input is of size $240 \times 240 \times 3$). The EfficientNet model has 7.8 million parameters to train, and thankfully this work has already been done for us. We will use this pre-trained network to find volcanoes[3] on Venus.

---

[3]The Oxford English Dictionary indicates that both volcanoes and volcanos are acceptable plurals. Based on the quotations provided therein, volcanoes seem to be more common in modern usage.

Fig. 6.8 Intermediate results from the convolution layers in the fashion MNIST model for two different inputs. The original image is shown, followed by the 64 convolutions from the first layer, max pooling with stride 2, 32 convolutions, and another max pooling of stride 2

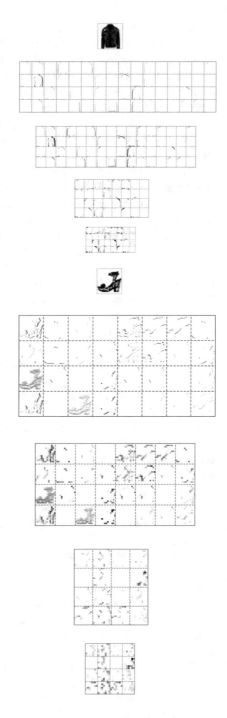

Taking a model that has been fit for one task and then adapting it for another task are called transfer learning. The idea behind transfer learning is that for complicated networks one requires a large amount of data and computational power to fit, as in EfficientNet. However, if we change the network slightly, such as adding an additional layer before the output, we can use a small amount of data to train that layer to learn how to adjust the result of the original network to the new task. This idea has successfully been applied in scientific applications to transfer learn the result of an experiment from a neural network that is trained on low fidelity simulations [4].

The data set that we use is based on the images collected by radar on the Magellan spacecraft from 1990–1994 [5]. The images are each of size 1024×1024 with a single gray color channel (i.e., the images are of size $1024 \times 1024 \times 1$) [6]. Each image comes with a key specifying the location in the image of a volcano and the confidence that the object is actually a volcano with the following meaning: 1 indicates a definite volcano, 2 is for a probably volcano, 3 is a possible volcano, and 4 indicates only a pit is visible. Given that no human has, or for matter could have given current technology, visited Venus to observe the volcanoes, this labeling does contain uncertainty. As provided by the UCI machine learning repository [7], there are 134 images.

To turn this into a useable data set for EfficientNet we take random 240×240 samples from the available images by randomly selecting a corner pixel for the 240×240 box. This provides the input data for our model. To get the dependent variables we use the original data to determine if there is a volcano in the sampled image and label the image with the lowest value volcano in the image (e.g., if there is a "1" volcano in the image, the dependent variable for that image is 1, regardless if there are less probable volcanoes in the image). If there is no volcano in the image, its label is set to 5. We created 11 596 images from the original data set and used an 80/20 training/test split. The classes in the set have roughly one-third of the images having no volcano and one-sixth of the images having each of the four classes of volcano probabilities.

This created data set is of size $240 \times 240 \times 1$, whereas EfficientNet expects a color image with 3 channels. We make data fit the expected input by simply copying the image data into three identical channels. Additionally, EfficientNet outputs a vector of size 1000 giving the probability of the image being each of 1000 classes. Our classification problem only has 5 classes for the volcano probability. To deal with this we add a softmax layer containing 5 nodes that takes as input the 1000 EfficientNet outputs. Therefore, our neural network will take as input a $240 \times 240 \times 3$ image that is fed into EfficientNet and the outputs from EfficientNet are passed through a softmax layer to get the probability of the image belonging to each of the five classes.

The softmax layer has a weight connecting each of the 5 nodes to the 1000 EfficientNet outputs. Including a bias for each node results in 5005 parameters in this final layer. When we train the model, we only train these 5005 parameters and leave the pre-trained EfficientNet network's parameters as static. This is the transfer learning step: we transfer the output of EfficientNet to solve a different problem by learning a small number of additional parameters. This allows us to leverage the work that went into designing and fitting EfficientNet to solve our problem. As a result our model learns how images from Venus activate the EfficientNet model and how to convert those results into our desired task.

Training the network for 100 epochs with a batch size of 32, we obtained the following results on the test data. The model predicted the correct class 52.4% of the time, with the true class being in one of top two classes predicted by the model in 75.6% of the cases. These accuracies can be compared to the random chance of guessing the correct label of 20% or a "top two" accuracy of 40%. Furthermore, the model indicates correctly whether there is volcano or not in the image 78.8% of the time.

These accuracies are notable for this "real-world" data set. There are missing values, such as areas where there is no data from the radar, as well as varying background darknesses in the image. These idiosyncrasies in the data were not corrected in the training data.

Example images and predictions are shown in Fig. 6.9. In the figure the six images with the highest predicted probability from the neural network for each class of volcano confidence are shown. From the image we see that the radar data does have missing values as well as artifacts from the radar collection process. Moreover, it appears that the size of volcanoes varies greatly, and yet the network was able to learn to identify these features.

Of course the network is not perfect. In Fig. 6.10 we show the worst predictions from the model: these had high confidences (predicted probabilities) for the wrong class. It seems that the network can be fooled by circular features in the data that are not volcanoes.

This example demonstrates the power of transfer learning: we can take a pre-trained model that required a vast amount of resources to train and augment it in a simple way. In this example we see that even the simple application of the principle can be powerful.

Transfer Learning

When one has a network that is trained to solve one task, it can be applied to another, similar task with a smaller amount of data by adding layers to the original network and training the model to learn a correction to the original model's outputs. This is called transfer learning, and because it is only training an addition to the original network, it allows one to apply sophisticated models with a large number of parameters requiring a large amount of data to train to be applied to new problems with only a small amount of data.

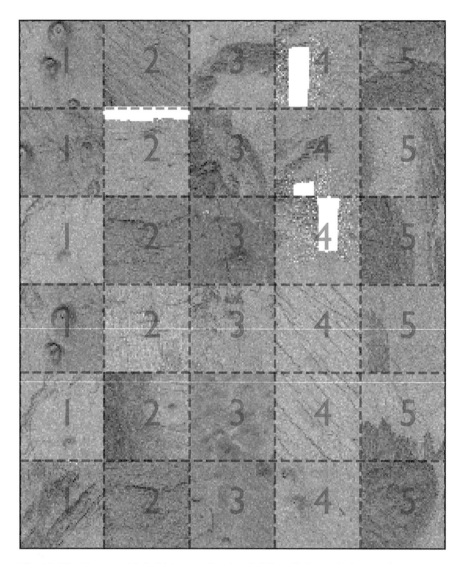

Fig. 6.9 The 6 images with the highest predicted probability of being each class are shown in each column with the first column being images the neural network identified as being in class 1. The numbers in each image are the correct class

Notes and Further Reading

The topic of convolutional neural networks is a rapidly evolving field with new tweaks on the concept being adding continually. One aspect of these models that has shown promise is residual layers where the inputs to a layer are operated on by a convolution and then added back to the inputs [8]. These ideas have been

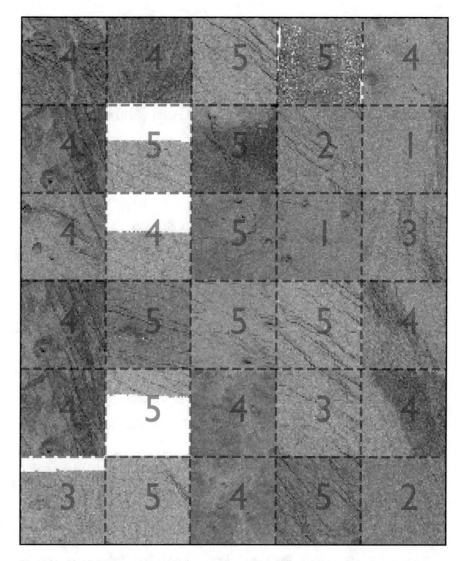

Fig. 6.10 The 6 images with the highest predicted probability of being in a class but with the wrong prediction are shown in each column with the first column being images the neural network identified as being in class 1. The numbers in each image are the correct class

shown to make more readily trainable networks that have high accuracy. A complete discussion of these networks is outside the scope of our study, but via transfer learning these models can be applied to a problem of the reader's interest.

Problems

6.1 Produce values of the function $\sin(t)$ at 100 equally spaced points from 0 to 2π. Add random noise to each point, where the noise at each point is a random number between -0.1 and 0.1. Call these values **f**.

- Compute the convolution of **f** with **g** as defined by Eq. (6.16) and plot the result versus the sine function that generated **f**.
- Compute the convolution of **f** with $\mathbf{g}_{D2}$ using $2\pi/100$ for h and plot the result versus $-\sin(t)$.

6.2 Construct a matrix of size 100×100 of random numbers between 0 and 1. Call this matrix **F**.

- Define a matrix **G** of size 3×3 where the matrix elements are

$$G_{ij} = \exp(-(i-2)^2 - (j-2)^2) \qquad i, j = 1, 2, 3.$$

- Take the convolution $\mathbf{F} * \mathbf{G}$ and plot it as a 2-D grayscale image, and compare this to the original matrix **F**.

References

1. Margaret Maynard-Reid. Fashion-MNIST with tf.Keras. https://medium.com/tensorflow/hello-deep-learning-fashion-mnist-with-keras-50fcff8cd74a.
2. Mingxing Tan and Quoc V Le. EfficientNet: Rethinking model scaling for convolutional neural networks. *arXiv preprint arXiv:1905.11946*, 2019.
3. Jia Deng, Wei Dong, Richard Socher, Li-Jia Li, Kai Li, and Li Fei-Fei. Imagenet: A large-scale hierarchical image database. In *2009 IEEE Conference on Computer Vision and Pattern Recognition*, pages 248–255. IEEE, 2009.
4. Kelli D Humbird, Jayson Luc Peterson, B. K. Spears, and Ryan G. McClarren. Transfer learning to model inertial confinement fusion experiments. *IEEE Transactions on Plasma Science*, 2019.
5. Gordon H Pettengill, Peter G Ford, William TK Johnson, R Keith Raney, and Laurence A Soderblom. Magellan: Radar performance and data products. *Science*, 252(5003):260–265, 1991.
6. Michael C Burl, Lars Asker, Padhraic Smyth, Usama Fayyad, Pietro Perona, Larry Crumpler, and Jayne Aubele. Learning to recognize volcanoes on Venus. *Machine Learning*, 30(2-3):165–194, 1998.
7. Dheeru Dua and Casey Graff. UCI Machine Learning Repository, 2017.
8. Kaiming He, Xiangyu Zhang, Shaoqing Ren, and Jian Sun. Deep residual learning for image recognition. In *Proceedings of the IEEE Conference on Computer Vision and Pattern Recognition*, pages 770–778, 2016.

Part III
Advanced Topics

Chapter 7
Recurrent Neural Networks for Time Series Data

I drove around for hours, I drove around for days
I drove around for months and years and never went no place

—Modest Mouse *Interstate 8*

Abstract In this chapter, we discuss neural networks that are specifically tailored to solve problems where the inputs are time series data, i.e., inputs that represent a sequence of values. We begin with the simplest such network, the basic recurrent neural network, where the output of a neuron can be an input to itself. We discuss how these networks function and demonstrate some issues with such a construction, including vanishing gradients when the input sequences are long. We then develop a more sophisticated network, the long short-term memory (LSTM) network to deal with longer sequences of data. Examples include predicting the frequency and shift of a signal and predicting the behavior of a cart-mounted pendulum

Keywords Recurrent neural networks (RNN) · Vanishing gradients; Long Short-Term Memory (LSTM) · Gated recurrent unit (GRU)

7.1 Basic Recurrent Neural Networks

In this chapter, we build machine learning models for prediction problems where the dependent variable is influenced by independent variables *and* the value of the dependent variable at previous times. A scenario where the data set changes over time is called a time series. Time series prediction problems are difficult because they usually require combining information about the history of the dependent variable with other independent variables. For instance, if we considered a time series of the temperature at a given location, the temperature 1 h later is likely influenced by the temperature right now. However, if other variables change, such as it starts raining or the sun sets, these can have a large effect on the temperature in 1 h.

© Springer Nature Switzerland AG 2021
R. G. McClarren, *Machine Learning for Engineers*,
https://doi.org/10.1007/978-3-030-70388-2_7

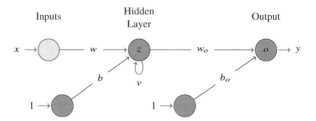

Fig. 7.1 Visualization of a basic recurrent neural network. Note that the output of z acts as an input to z with weight v

Time series prediction problems need to deal with the historical behavior of the function as well as quantify the influence of other variables. We accomplish this by making the neural network have recursion: that is, the output of a neuron serves as an input to itself. This enables the network learn how to consider the past as well as other variables in the model. There are several types of recurrent neural networks, and we begin with the simple recurrent neural network.

In its simplest form, a recurrent neural network (RNN) consists of a single neuron where the neuron output serves as an input. If z_t is the output of the neuron at time t, and x_t is an independent variable at time t, then the neuron would take a value of

$$z_t = \sigma(wx_t + vz_{t-1} + b), \tag{7.1}$$

where w and v are weights, b is a bias, and $\sigma(u)$ is an activation function. If we embed this neuron in a network with a single hidden layer, we can visualize it as in Fig. 7.1. In this figure we show a RNN that takes a single input x and produces an output y. This neural network computes the output y as

$$y_t = o(w_o z_t + b_0), \tag{7.2}$$

where $o(u)$ is the output activation function, w_o is a weight, and b_o is a bias. This RNN has 5 parameters that need to be trained: w, v, b, w_o, and b_o.

Another way to picture a RNN is to "unroll" it. That is to show the recursion by repeating the neural network at each time level, as is shown in Fig. 7.2. In this depiction, we can directly see the connections between the time levels. Also, we can see that z_t contains some information about the history of the time series. The neural network learns what information about the history of the time series should be encoded in the state variable z_t so that when combined with x_{t+1} via the weights and biases, the correct value of y_{t+1} can be calculated. Notice that if the weight v is zero, then this is just a standard feed-forward network.

Of course, it is possible to have more than one recurrent node in a neural network. Adding further recursion allows the network to have several different variables store information about the history of the time series. Adding more recurrent nodes is

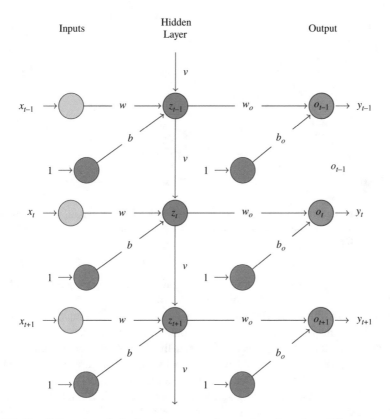

Fig. 7.2 Unrolled version of a basic recurrent neural network to show three different times and the connections between them. Note that the weights do not change with time

akin to making the network deeper in the time domain. There are now connections between the different recursive nodes in the network. A RNN with two recursive nodes, a single input, and a single output is shown in Fig. 7.3. In the figure, we can see that the network gets more difficult to visualize as there are recursive connections and connections between the recursive nodes at different time levels. Mathematically, the values of the recursive nodes are

$$z_t^{(1)} = \sigma(w_{0\to 1}x + v_{1\to 1}z_{t-1}^{(1)} + v_{2\to 1}z_{t-1}^{(2)} + b_1), \tag{7.3}$$

$$z_t^{(2)} = \sigma(w_{0\to 2}x + v_{2\to 2}z_{t-1}^{(2)} + v_{1\to 2}z_{t-1}^{(1)} + b_2).$$

The fact that this network has two recursive nodes and they are connected to each other gives the network the ability to encode two pieces of information about the time history of the data, but also allows that information to interact with each

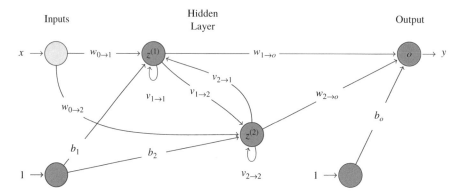

Fig. 7.3 Visualization of a recurrent neural network with a single input, x, and two recursive nodes, $z^{(1)}$ and $z^{(2)}$. In this case, each of the recursive nodes takes as input both values of z from the previous time

other. This is indeed a very flexible network, and it becomes more flexible (and more complicated to draw) as we add more recursive nodes.

It is also possible to create neural networks where recursive nodes are a piece of a much larger network. For instance, we could have a deep feed-forward network or a convolutional neural network between the inputs and the recursive nodes. Additionally, we can have hidden layers between the recursive nodes and the output layer, for instance, a softmax layer to solve a classification problem. These additional levels of complexity give the network further flexibility in the types of relations that the network can model at the potential cost of having a more difficult model to train.

> **Recurrent Neural Networks**
>
> A recurrent neural network (RNN) is a network where the outputs of some neurons are inputs to themselves. This recursion allows the values from previous evaluations of the model to affect the future. This feature is useful for time series data where the independent variables as well as the current value of the dependent variable can influence future values.

7.1.1 Training RNNs and Vanishing Gradients

Training recursive neural networks can be done using back propagation in the same way that feed-forward and convolutional neural networks are trained. In the typical training scenario, a single training instance will have T values for x_t, i.e., $(x_1, x_2, \ldots, x_T)$, and a known value for y_T. In this case, we train the network with the inputs from T times and have it produce a prediction for the dependent variable at time T. If we unroll the RNN T times, then we can perform back propagation from y_T to get the derivative of the error with respect to each of the parameters.

To see how this works, we consider a single training point for the neural network shown in Figs. 7.1 and 7.2 with linear activation functions for $\sigma(u)$ and $o(u)$. If we consider the derivative of y_T with respect to v, we get

$$\frac{\partial y_T}{\partial v} = w_o \frac{\partial z_T}{\partial v} \tag{7.4}$$

$$= w_0 \frac{\partial}{\partial v}(v z_{T-1} + w x_T + b)$$

$$= w_o \left(v \frac{\partial z_{T-1}}{\partial v} + z_{T-1} \right)$$

$$= w_0 \left[v \frac{\partial}{\partial v}(v z_{T-2} + w x_{T-1} + b) + (v z_{T-2} + w x_{T-1} + b) \right]$$

$$= w_o \left[v \left[v \frac{\partial}{\partial v}(v z_{T-3} + w x_{T-2} + b) + (v z_{T-3} + w x_{T-2} + b) \right] + (v z_{T-2} + w x_{T-1} + b) \right]$$

$$\vdots$$

To terminate this sequence, we set $z_0 = 0$. Inspecting the equation, we see that if $T = 2$ then

$$\frac{\partial y_2}{\partial v} = w_o(w x_1 + b). \tag{7.5}$$

Similarly, if $T = 3$ we have

$$\frac{\partial y_3}{\partial v} = w_o(2v(b + w x_1) + b + w x_2); \tag{7.6}$$

and $T = 4$ gives

$$\frac{\partial y_4}{\partial v} = w_o \left(3b v^2 + 2b v + b + 3v^2 w x_1 + 2v w x_2 + w x_3 \right). \tag{7.7}$$

We need not continue further. What we observe here is the derivative of the output with respect to v will have terms containing v^{T-2}. Therefore, as T gets large, that is we have long time series, the derivative will have a large power of v in it. This means that if $|v| > 1$, then the magnitude of the derivative goes to infinity as T gets large. Similarly, if $|v| < 1$ then v^{T-2} goes to zero as T gets large. Typically, when we train a neural network we start with values of the weights that are close to zero. Therefore, when we evaluate the derivative of y with respect to v in training the network, the influence of data at early times is functionally zero when T is large. This is known as the vanishing gradients problem. It makes it very difficult for the network to learn to propagate information from the time series from early times to later times because the early times have little influence on the late time data when T is large. We will discuss approaches to deal with the vanishing gradients problem later in the chapter.

Vanishing Gradients

One drawback of RNNs is that the recursion behaves as though we have added hidden layers to the network. Therefore, when using back propagation to compute derivatives, the problem can arise that values early in the time series cannot affect the later time values due to the phenomenon of vanishing gradients. This causes RNN models that deal with time series at a large number of points to be difficult to train.

7.1.2 Example RNN: Finding the Frequency and Shift of an Oscillating Signal

To demonstrate the properties of RNNs, we use a simple problem to demonstrate their behavior. We will feed the network signals of the form

$$y(t) = \sin(at + b) + \epsilon(t), \tag{7.8}$$

where a and b are unknown constants for a given signal chosen from a standard normal distribution, and $\epsilon(t)$ is a noise term where for each time the value is drawn from a normal distribution with mean 0 and standard deviation of 0.01. To train the RNN, we create $2^{20} \approx 10^6$ signals sampled at 10 equally spaced points between 0 and 2π with a spacing between points of $\Delta = \pi/9$. This training data creates a matrix $\mathbf{X}$ of size $2^{20} \times 10$. The dependent variable for this problem is a matrix $\mathbf{Y}$ of size $2^{20} \times 1$, which contains the value of $y(2\pi + \Delta t)$ without the noise term added. We also construct a test data set containing 2^{18} series, about one-quarter of the training data.

The goal here is to train a model that can predict the next value of the series by only having information about the current and past states of the signal. To do this, it must learn the frequency of the signal, represented in the constant a, and the shift in the signal given by b. The first model we build has identical structure to that in Fig. 7.1: It has a single recursive unit followed by a single output neuron. This network has 5 parameters that need to be trained. We train the model using 200 epochs and a batch size of 256. After training, the network has an RMS error of about 0.3. It seems that this RNN is too simple to learn the behavior of these signals.

To better accomplish the task of this example, we try a more complex RNN. This model has 3 recurrent neurons and a dense layer of 3 neurons after the recurrent neurons. This network has similar structure to that of Fig. 7.3 except that there are 3 recurrent neurons (instead of 2) and there are 3 neurons between the output and the recurrent layer. This network performs better than the simple network with an RMS error of 0.109 on the test set.

These networks were trained to predict the eleventh point in a sequence, but we could use them to generate more points of the signal. For instance, we can use the prediction of the model and last 9 points to predict the twelfth point in the sequence. This could be repeated many times and compared with the true value of a signal. We do this type of repeated prediction to see how the model behaves in extrapolating from what it has seen into a continued signal. We use the model to predict the next 10 points in the sequence. In practice, this means that we pick values of a and b and produce a signal at 10 points. We then feed these 10 points into the model to predict point 11. Then points 2-11 are used to predict point 12 and repeat until we have predicted 10 points.

The results from this prediction exercise for a particular case are shown in Fig. 7.4. In the figure, only the 10 predicted points from each RNN are plotted and the true signal is also shown. We can see that the first prediction is somewhat close to the correct value for both models, though it is closer for the more complicated RNN model. As more predictions are made using the prediction from the network at previous times, we see that the errors compound. The complex network does a better job and has the predicted signal have the shape of a sine wave (without the correct amplitude), but the simple network does not produce results that correctly capture the features of the true signal.

As noted above, RNNs cannot learn information from the distant past due to the phenomenon of vanishing gradients. To demonstrate this, we retrain the more

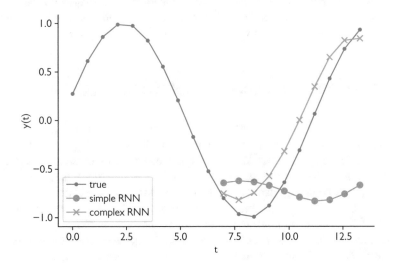

Fig. 7.4 Comparison of a simple RNN with one recurrent neuron and a more complex RNN with 3 recurrent neurons and a dense layer of 3 neurons on predicting the next 10 points in a sequence

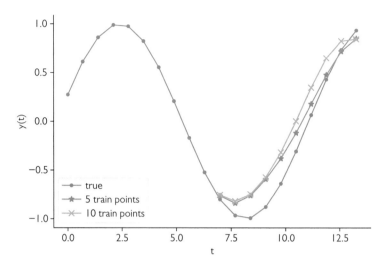

Fig. 7.5 Comparison of the more complicated RNN using 5 or 10 training points per signal on the task of predicting the next 10 points

complex network with only 5 values from the true signal. That is, our training data only contains half the number of values for each signal. The phenomenon of vanishing gradients indicates that this network should have similar performance to the network trained with 10 points per signal because the ability to predict a value should not depend on values from the signal 10 time levels in the past. Indeed, this is what we see. In terms of RMS error, the network using only 5 points in the training has an error of size 0.0989. Furthermore, comparing the networks on the task of predicting the next ten values shows that the network trained with 5 points performs slightly better than that trained with 10 points, as shown in Fig. 7.5.

7.2   Long Short-Term Memory (LSTM) Networks

The basic RNN has the problem that via vanishing gradients it is difficult to carry information from several steps in the past. To address this issue, long short-term memory (LSTM) [1, 2] networks were developed. This network adds an additional path of recursion through the network called the "state" that is designed to be passed over time without being changed much. The network uses the state as well as the output for the dependent variable in the recursive process so that the network can, in a sense, keep different kinds of information propagating in time.

To demonstrate the LSTM, we will build up the network one step at a time for a network with a single state and output. In this way, we can understand all the connections in the network and the operations that are occurring. We begin with the network knowing the state, C_{t-1}, and output, z_{t-1}, from the previous step.

Fig. 7.6 Visualization of an LSTM network showing the forget gate in Step 1. The node with "×" inside denotes the elementwise multiplication of the inputs to the node. The biases are not shown in this or in subsequent LSTM figures for clarity

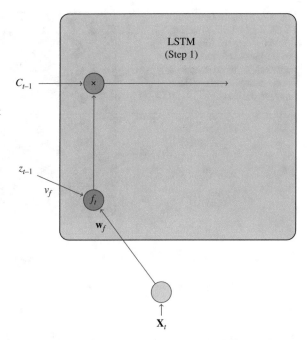

Step 1: Decide What to Forget

Knowing the previous state and output we compute a value between 0 and 1 that combines the inputs at x_t with the previous output to get a number between 0 and 1:

$$f_t = \sigma_f(\mathbf{w}_f \cdot \mathbf{x}_t + v_f z_{t-1} + b_f), \tag{7.9}$$

where $\mathbf{w}_f$ is a vector of weights for each of the J inputs in $\mathbf{x}_t$, v_f is a weight, b_f is a bias, and $\sigma_f(u)$ is a sigmoid function giving a value between 0 and 1. We call f_t the *forget gate value*. This number is multiplied by the current state to get the value of the state after forgetting, C_{t-1}^f,

$$C_{t-1}^f = f_t C_{t-1}. \tag{7.10}$$

Clearly, if $f = 0$ the state goes to zero and if $f = 1$ the state is unchanged.

This first step of the LSTM network is shown in Fig. 7.6. Notice that the input and the previous output, z_{t-1}, only influence the state via the forget gate value.

Step 2: Propose a New State

Now we combine the output from the previous step z_{t-1} with the inputs to propose a new state. The new proposed state, $\tilde{C}_t$, is the output of a neuron combining the inputs and z_{t-1}:

$$\tilde{C}_t = \sigma(\mathbf{w}_p \cdot \mathbf{x}_t + v_p z_{t-1} + b_p), \tag{7.11}$$

Fig. 7.7 Visualization of an
LSTM network showing the
first two steps: the forget gate
and the proposed state. The
node with $+$ inside denotes
the elementwise sum of the
inputs to the node

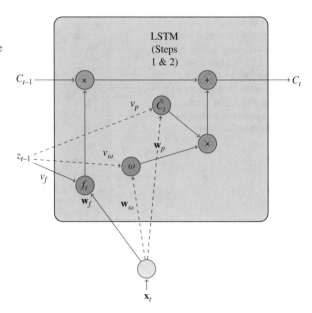

where the subscript p denotes parameters used in the proposal of a new state. We
also denote a sigmoid function σ as a function taking values between -1 and 1 so
that the state can be positive or negative. We then decide how much to weigh this
state when adding it to C_{t-1}^f to get the new state. We call this weight ω and calculate
it as

$$\omega = \sigma_f(\mathbf{w}_\omega \cdot \mathbf{x}_t + v_\omega z_{t-1} + b_\omega). \tag{7.12}$$

Once again we have used σ_f to assure that the weight ω is between 0 and 1. Using
this weight we create the new state by adding the weighted, proposed state to the
previous state after forgetting

$$C_t = C_{t-1}^f + \omega \tilde{C}_t. \tag{7.13}$$

The calculations in step 2 are added to the LSTM schematic in Fig. 7.7. In this
figure, we have denoted some of the connections with dashed lines for clarity. In
this step, we produce the new value for C_t. Here $\mathbf{x}_t$ and z_{t-1} compute a weight, ω
that is combined via elementwise multiplication with the proposed state, $\tilde{C}_t$, which
is also a function of $\mathbf{x}_t$ and z_{t-1}. The proposed state is then added to the result from
the forget gate.

Fig. 7.8 Visualization of a complete LSTM network at time t

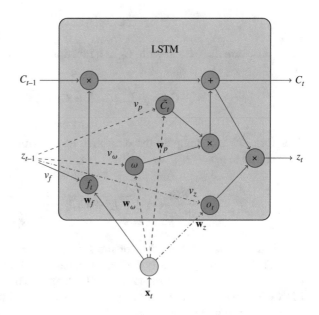

Step 3: Use the New State to Get the New Output

Finally, we use the new state, C_t, and the previous output, and the inputs to get the new output. The new output, z_t, is the product of a standard neuron times an activation function applied to the state:

$$z_t = o_t \sigma(C_t), \qquad o_t = \sigma_f(\mathbf{w}_z \cdot \mathbf{x}_t + v_z z_{t-1} + b_z). \tag{7.14}$$

In Fig. 7.8 we can see the entire LSTM. The state only influences the future state as well as the output value of z_t. This is in contrast to the inputs $\mathbf{x}_t$ and the output from the previous step z_{t-1}: These influence the forget gate, the proposed new state, and the output.

The number of parameters in the LSTM model is $4J + 4 + 4 = 4(J + 2)$. The $4J$ comes from the weight vectors in the forget gate, the proposal for the new state, the weighting of the proposed state, and in the calculation of the new output. The other two terms come from the 4 weights on z_{t-1} and the four biases.

7.2.1 How LSTMs Get Around the Vanishing Gradients Problem

In introducing the state, the LSTM network removes the vanishing gradients problem. That is, the state can be largely unchanged by the network from time level to time level without the value of the state going to zero exponentially with the number of time levels. To see this, we consider a network where the weights are such that $\omega = 0$. In this case, the value of C_t will be

$$C_t = f_t C_{t-1}. \tag{7.15}$$

Using this formula, we can write C_t as a function of C_{t-2} or C_{t-k} as

$$C_t = f_t f_{t-1} C_{t-2} = C_{t-k} \prod_{k'=0}^{k-1} f_{k'}. \tag{7.16}$$

Therefore, we can compute the derivative of C_t with respect to C_{t-k} using the product rule to get

$$\frac{\partial C_t}{\partial C_{t-k}} = \prod_{k'=0}^{k-1} f_{k'} + C_{t-k} \frac{\partial}{\partial C_{t-k}} \prod_{k'=0}^{k-1} f_{k'}. \tag{7.17}$$

This first term is just a product of the k forget gates between time levels t and $t-k$. These are not necessarily the same number, therefore, it is possible to train the network so that this term does not vanish for large k. This is in contrast to the basic RNN where powers of a weight v appear so that terms of the form v^k go to zero or to an infinite magnitude as k gets large.

Long Short-Term Memory (LSTM)

Long short-term memory units contain two pieces of information at each time level: an output and a state. LSTMs have a forget gate that decides, based on the previous output and the independent variables, whether to forget the current state and proposes a new state. Then the new state is combined with the output from the previous step and the independent variables to get a new output. The combination of the form of the forget gate and the way the output is updated allows LSTMs to avoid the vanishing gradients problem.

7.2.2 Variants to the LSTM Network

In the description above, there was a single LSTM "unit" in the network. In principle there could be several units, each with its own state and output. In this case, the inputs to the network would be modified so that $\mathbf{z}_t$ and $\mathbf{C}_t$ would be vectors of the same length as the number of units. This would require a vector of weights $\mathbf{v}$ for each operation involving $\mathbf{z}$, e.g. in the forget gate $\mathbf{v}_f$. The other operations would be executed elementwise, i.e., $\sigma(\mathbf{C}_t)$ would be a vector of σ applied to each entry in $\mathbf{C}_t$. When the number of units increases to N units, the number of free parameters is then $4(JN + N^2 + N)$.

As with the basic RNN, we can connect other layers between the independent variables and the LSTM units and/or place layers between the outputs z_t and the actual dependent variables. Additional layers added after the LSTM module will not typically take the state, C_t, as an input.

There are two important variants of LSTM that we mention in passing. The first is the peephole LSTM [3] where the C_{t-1} is also an input to the forget gate and the weight ω, and the value of the new state C_t is an input to the neuron in Eq. (7.14). This allows the terms that construct the state and the output to have knowledge of the state or to peek at the state and use it to influence the results. The other variant is the gated recurrent unit (GRU) [4], which acts in many ways like an LSTM unit, but only has a single output rather than adding the state. This output is recursive in a way that uses a weight similar to the forget gate so that information can be propagated for long times. Because the GRU does not add the extra degree of freedom of the state, these networks are typically simpler than LSTM models.

7.2.3 Example LSTM: Finding the Frequency and Shift of an Oscillating Signal

We can apply LSTM models to the problem of estimating the functions of the form of Eq. (7.8). We train a model that takes 10 data points as input and predicts the next 10 points. In this case, the LSTM model can be thought of as having 10 outputs to predict. The model we build has a layer with 16 LSTM units. From these 16 units, the $\mathbf{z}_t$ are fed into a fully connected layer of 10 hidden neurons with the ReLU activation function. These 10 nodes then connect to an output layer of 10 nodes, one each for the 10 output values. We train this network using the same data as in the RNN example. After training the model, we find that the mean-absolute error is 0.0323 on the training data and 0.0331 on the test data.

In Fig. 7.9 we show 3 validation cases for the LSTM model. These results indicate that model is able to correctly predict the behavior of the signal, including points where the function passes through a minimum, a property that the basic RNN models had difficulty with. We also fit another model, this time using 15 values from the sequence as input and predicting the next 5 values in the sequence as output. RNN networks would be expected to do worse on this problem because of the vanishing gradients problem. In the case of the LSTM, adding more points improves the model performance, as shown in Fig. 7.9b.

It is clear that a neural network with LSTM nodes is able to solve this problem. In future chapters, we will see that these techniques can be used in conjunction with other advanced techniques to create powerful models for science and engineering problems.

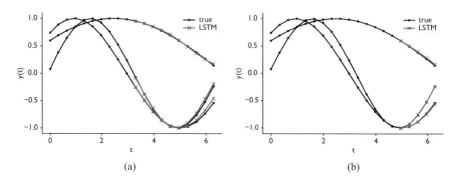

Fig. 7.9 Performance of the model with 16 LSTM units applied to problem of predicting the next values of a sequence given the first 10 or 15 values. (**a**) Length 10 input sequence; predict 10. (**b**) Length 15 input sequence; predict 5

7.3   Case Study: Determining the Behavior of a Cart-Mounted Pendulum

The system we want to model with recurrent neural networks is the cart-mounted pendulum. We have a mass m at the end of a rigid pendulum. The other end of the pendulum is attached to a cart of mass M that can move to the left or the right. The cart is on an elevated track so that the mass can swing below the cart. As the mass swings the cart moves as the center of mass of the system changes. This system is shown in Fig. 7.10.

The goal of our model will be to take data from the pendulum system for the position of the pivot point, x, and the angle of the pole θ, for a system with unknown masses, M and m, and length ℓ, and to predict the future behavior of the pendulum. To produce data to train our machine learning model, we will integrate the equations of motion for this system.

Equations of Motion for the Pendulum
As derived in [5], the equations for motion for the cart-mounted pendulum system are given by the second-order differential equations:

$$\ddot{x} = \frac{m \sin \theta \left(\ell \dot{\theta}^2 + g \cos \theta \right)}{M + m \sin^2 \theta}, \tag{7.18}$$

$$\ddot{\theta} = \frac{-m\ell\dot{\theta}^2 \cos \theta \sin \theta - (M + m)g \sin \theta}{\ell(M + m \sin^2 \theta)}. \tag{7.19}$$

In these equations, g is the acceleration due to gravity.

For our training data, we randomly select masses, length of the pendulum, and the acceleration due to gravity as well as random values for the initial values of θ, $\dot{\theta}$, x, and $\dot{x}$. With this information we use a differential equation integrator to

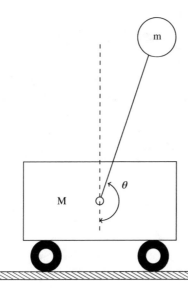

Fig. 7.10 Schematic for the inverted pendulum. A mass is connected to the cart by a rigid arm. The cart rolls freely as the mass We define the pendulum angle to be in the range $\theta \in [-\pi, \pi]$. The length of the pendulum is ℓ

update the position and velocity of the cart, $(x, \dot{x})$, the angle of the pendulum and its angular velocity, $(\theta, \dot{\theta})$, to get new values of the system state at time $t + \Delta t$. This allows us to virtually operate the system by specifying an applied force and seeing what happens to the cart and pendulum. For the masses and length, M, m and ℓ, we selected them from a uniform distribution between 0.5 and 1.5 (in units of kilograms or meters), the acceleration due to gravity g was sampled from a normal distribution with mean 9.81 m/s^2 and standard deviation of 0.005 m/s^2. The initial states had θ uniformly sampled between $-\pi$ and π, with x, $\dot{x}$, and $\dot{\theta}$ drawn from uniform distributions between -1 and 1.

For training data for our model, we sampled $2^{13} = 8192$ different systems and initial states. The state of each system was recorded at intervals of 0.1 s between $t = 0$ s and 10 s, for 100 data points per system. From these data, we created a training set by randomly selecting a system and a starting point in that system's time series. From that time point we select 60 values for the position x and angle θ using the first 40 values of each as inputs and the last 20 time points of x and θ as the dependent variables that we want our model to predict. We add to the dependent variables the values of m, M, ℓ, and g to make the total number of dependent variables $44 = 2 \times 20 + 4$. We sample $2^{21} = 2\,097\,152$ such time series and system parameters for training data (i.e. our training data has size $2^{21} \times 40 \times 2$ for inputs and $2^{21} \times 44$ for outputs) and 2^{19} time series as test data.

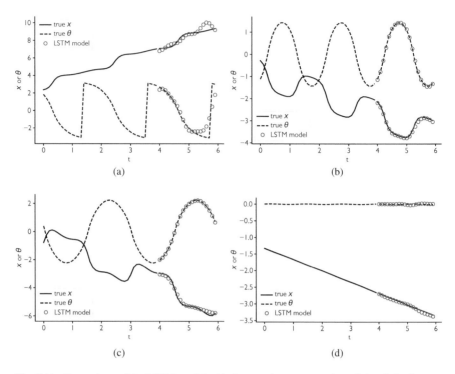

Fig. 7.11 Comparison of the LSTM model with the actual cart-mounted pendulum behavior as a function of time for four test cases, (**a**) through (**d**). The first 40 time points are inputs to the model, and the model predicts the next 20 points for both x and θ

7.3.1 Model Design and Results

The model we use consists of an LSTM layer with 16 units connected to another LSTM layer with 16 units. The second LSTM layer is connected to an output layer with 44 nodes corresponding to the 44 dependent variables (20 each for x and θ and the 4 parameters m, M, ℓ, and g). Before training, we normalize the input data by subtracting its mean and dividing by its standard deviation for each time point. We do the same normalization for the 44 dependent variables. After 40 epochs of training using the Adam optimizer, we obtain a mean-absolute error of 0.0714 on the training data and 0.0703 on the test data. In terms of the normalized data, this means that the model prediction is on average within about ± 0.07 standard deviations from the correct value.

In Fig. 7.11 results for 4 different test cases are shown. These cases span a variety of phenomena, and our model seems to be able to handle each of them (with varying degrees of success). In the case in Fig. 7.11a, the pendulum makes several full revolutions. This makes the value of θ discontinuous as it switches from $-\pi$ to π repeatedly. The model tries to smoothly match this transition, and does a reasonable

Fig. 7.12 Longer prediction with the LSTM model. The model is run 6 times to predict the final 120 points in the time series

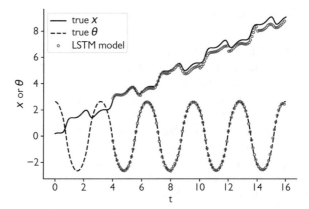

job, but the continuous transition in θ causes the model to predict incorrect response in x. In hindsight, perhaps we could have tried to predict $\cos \theta$, for example, to have a continuous variable. This does highlight how selection of input and dependent variables can affect the model performance.

At the opposite extreme, we have Fig. 7.11d where the pendulum does not oscillate much at all. In this case, the model correctly predicts that the position of the cart continues along the linear trajectory established in the input. The most common case in the training data is similar to the time series in Fig. 7.11b, c. These results have the pendulum oscillating from positive to negative angles and pulling the cart in one direction. For both of these cases, the model correctly predicts the path of the cart and the frequency of the oscillation.

Another way to test the model is to apply it repeatedly to predict longer time series. In this case, we feed the model 40 time points to predict the next 20 time points. We could then use the 20 predicted points (along with 20 of the original inputs) as inputs to the model to get time points 21 through 40. Repeating this process, we could make the time series as long as we wanted. We would expect a degradation of the model performance because the inputs to the subsequent model evaluations will have errors in them. We can observe this phenomenon in Fig. 7.12. The model can predict the overall behavior of the cart and pendulum, but as the time gets farther away from the true input data, the errors in the prediction compound to give noticeable discrepancy between the model and the true behavior. In this case, it might be better to apply the approach from Sect. 2.5 and directly estimate the equations of motion and use the derived model to extrapolate the behavior of the system.

Notes and Further Reading

The topic of RNNs and other methods for time series could be an entire book-length discussion in itself. There are techniques that originated in machine translation called sequence-to-sequence (seq2seq). These models train an encoder and decoder where the encoder takes the original sequence, converts it to a hidden representation much like the LSTM, and then the decoder network takes that hidden representation and converts it to the output sequence. The input and output sequences can be of arbitrary length, and this is one of the benefits of these approaches. One variant of the seq2seq approach that has found widespread use in machine translation is the attention network [6] that allows the network to find what parts of the inputs are most important to decoding a specific output. The discussion above of RNNs and LSTMs is a good foundation for further exploration in this area.

Problems

7.1 Construct a training data set of 10^5 examples from the logistic map

$$x_{n+1} = rx_n(1 - x_n),$$

by randomly selecting a starting point x_0 between 0 and 1, selecting r randomly between 0 and 4, and computing the next five terms in the sequence. Use these five points in a network using RNNs and a network using LSTMs to predict the sixth point. Repeat the exercise using 10 points from the sequence as inputs and try to predict the eleventh point. Does the model perform better for certain values of r?

7.2 Repeat the previous problem but now try to predict the next 5 and 10 points in the sequence using 5 and 10 input values respectively.

7.3 Consider the nonlinear differential equation [7]

$$m\frac{d^2y}{dt^2} + 200\frac{dy}{dt} + 10000(1 + 5000y^2)y = 10000\sin(2\pi vt),$$

with initial condition $y(0) = 0$ and $\dot{y}(0) = 0$. Using a numerical integrator generate 2000 examples from this equation at 1000 times between $t = 0$ and 1 with m randomly selected between 0.5 and 1.5 and v randomly selected between 15 and 25. From each of these examples, select 10^5 training cases where one of the 2000 examples is selected randomly and a start time is selected randomly. From the start time, collect 20 consecutive time points as the input and 10 time points as the output. Build an LSTM neural network to predict the output as a function of the input.

7.4 Recurrent neural networks can be used with image data as well. Using the Fashion MNIST data set, build an RNN or LSTM network that takes as input each row from the image starting at the top. That is, this network will take in 28 inputs, over 28 time steps, to have the entire image as input. The output of the network will be a softmax layer that identifies which class the image belongs to. This problem has the network reading in the image as though it were a time series to determine the class it belongs to.

References

1. Sepp Hochreiter and Jürgen Schmidhuber. Long short-term memory. *Neural computation*, 9(8):1735–1780, 1997.
2. Felix A Gers, Jürgen Schmidhuber, and Fred Cummins. Learning to forget: Continual prediction with LSTM. 1999.
3. Hasim Sak, Andrew W Senior, and Françoise Beaufays. Long short-term memory recurrent neural network architectures for large scale acoustic modeling. 2014.
4. Junyoung Chung, Caglar Gulcehre, KyungHyun Cho, and Yoshua Bengio. Empirical evaluation of gated recurrent neural networks on sequence modeling. *arXiv preprint arXiv:1412.3555*, 2014.
5. Russ Tedrake. Underactuated robotics: Learning, planning, and control for efficient and agile machines course notes for MIT 6.832. 2009.
6. Dzmitry Bahdanau, Kyunghyun Cho, and Yoshua Bengio. Neural machine translation by jointly learning to align and translate. *arXiv preprint arXiv:1409.0473*, 2014.
7. Joseph M Powers and Mihir Sen. *Mathematical Methods in Engineering*. Cambridge University Press, 2015.

Chapter 8
Unsupervised Learning with Neural Networks: Autoencoders

Along the lips of the mystic portal he discovered writings which after a little study he was able to decipher.

—Nathanael West *The Dream Life of Balso Snell*

Abstract In this chapter, we study an approach to unsupervised learning using neural networks called an autoencoder. An autoencoder attempts to find a small set of latent variables, called codes, that can be used to represent a larger data set. This is similar to the singular value decomposition we previously discussed where a small set of uncorrelated variables is sought to represent a larger set of variables. In an autoencoder, we fit a neural network of the identity: $f(x) = x$, where the hidden layers have an hourglass structure such that a central hidden layer has a smaller number of nodes than the inputs. The values of these central hidden units give us the latent variables. There are versions of autoencoders that use convolutions that we discuss as well. We apply autoencoders to hyper-spectral data from foliage, as well as image data from a set of physics simulations. Finally, we demonstrate that the latent variables found by an autoencoder can be used as inputs to supervised learning problems.

Keywords Autoencoders · Data compression · Convolutional autoencoders · Latent space · Encoder · Decoder

8.1 Fully Connected Autoencoder

To this point we have only discussed neural networks for supervised learning problems where we have a target dependent variable. In this chapter, we look at how we can use neural networks for unsupervised learning problems to find reduced representations of data or other structure. We discussed unsupervised learning before in Chap. 4. In this chapter, we extend those ideas using neural networks to get more powerful, though less explainable, methods. By less explainable, we mean that the results from a neural network-based method may not have the direct

© Springer Nature Switzerland AG 2021
R. G. McClarren, *Machine Learning for Engineers*,
https://doi.org/10.1007/978-3-030-70388-2_8

explanation that, for instance, the singular value decomposition has in terms of linear combinations of variables. Nevertheless, the resulting reduction in the size of the data can be impressive. We begin with a nonlinear analog to the singular value decomposition: the autoencoder.

The idea of an autoencoder is simple, we try to learn an identity mapping, i.e., a function that returns its argument: $\mathbf{f}(\mathbf{x}) = \mathbf{x}$, except the function has reduced the complexity of the input data in some form. For instance, if we consider $\mathbf{x}$ having N components, a neural network that had a single hidden layer with $N/2$ nodes, and N output nodes would have effectively reduced the data by nearly factor of 2.

This network: N inputs, $N/2$ hidden units, and N output nodes is visualized in Fig. 8.1. This figure shows a very simple autoencoder. The first part of the network (going from N inputs to the $N/2$ hidden units) is called the encoding layer or encoder. The second half (going from the $N/2$ hidden units to the N output nodes) is called the decoder. When we train this network, we adjust the weights and biases in the network so that the encoded values, that is, the values of the hidden units, give back the original data when decoded. In this network, there are $(N/2+1)N$ weight and bias parameters in each of the encoding and decoding layers.

We call the values of nodes in the middle layer, $\mathbf{z}(\mathbf{x})$, the latent variables or the codes. The basic idea is that if such a network can be trained we can store only the codes for our data, and use the decoder when we need the full representation of the data. In this sense, we have compressed the input data to the codes. After we train such an autoencoder, assuming the error in the map $\mathbf{f}(\mathbf{x}) = \mathbf{x}$, is low, then we only need to store the $N/2$ encoded values for our data $\mathbf{x}$ and can use the decoding layer to reproduce the full vector when needed. Therefore, if we have M vectors of data we need to store $M * N/2$ numbers, that is, the codes for each case, plus the $(N/2 + 1)N$ weights and biases in the decoding layer. If $M \gg N$, then we have effectively reduced our storage requirement in half.

Of course, if a single hidden layer can be used as an autoencoder, we could add further hidden layers to the network to try to get better compression. These networks typically have an hourglass shape. In the encoding layers the width, or number of nodes in a layer, reduces as one gets deeper. A typical pattern would be N input nodes, $N/2$ nodes in the first hidden layer, $N/4$ in the subsequent hidden layer, and a reduction by a factor of two in each additional layer. In the center of the network will be a layer that contains the encoded variables. The decoder layer is typically symmetric about the encoding layer with more nodes being added as one moves from the encoded variables to the N output nodes. For instance, the decoding layer could double the number of hidden units in each subsequent layer. There is no requirement that the network must be symmetric or that there needs to be a constant factor reduction/increase between the layers, but it has been observed that these simple structures seem to work in practice.

A deep autoencoder with this symmetric structure is shown in Fig. 8.2. The network reduces in the width of the hidden layers in the encoder until it reaches the encoded variables that are the values of $z_k^{(2)}(\mathbf{x})$. The network then expands back

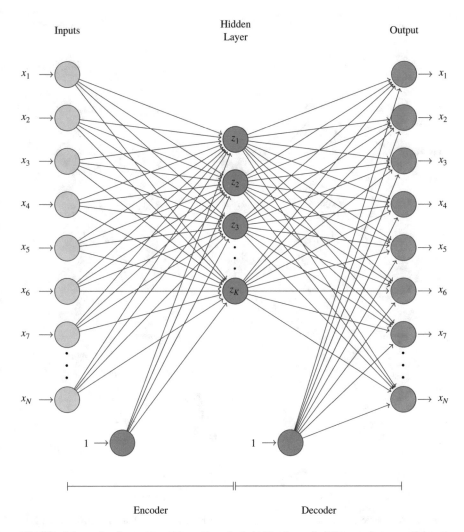

Fig. 8.1 Schematic of an autoencoder with a single hidden layer. In this case, there are N inputs that are encoded into K degrees of freedom in the hidden layer, which are then decoded back into the N original inputs

to the size of the original data in the decoder. Though the network is symmetric, the weights in the encoder and decoder are not necessarily the same.

Once we have built an autoencoder, we can use the encoded variables to gain insight into the structure of the data. For example, we could look at those cases that have extreme values in each of the encoded variables. This can give us an idea of what the encoded variables mean such as if they are encoding particular features of the data. We must be careful, however, because there are nonlinearities in the encoding/decoding process that could make that inference difficult.

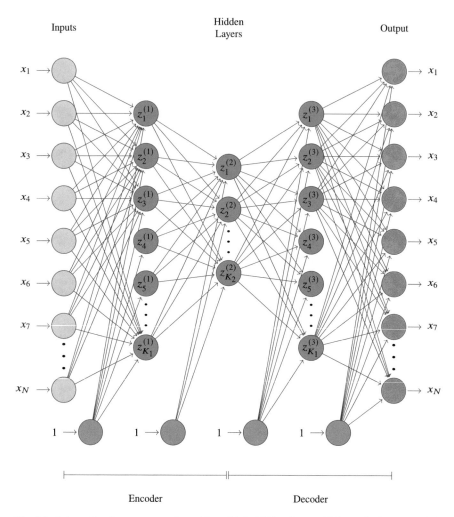

Fig. 8.2 Schematic of an autoencoder with multiple hidden layers. This particular network is symmetric about the encoded variables in that the encoder and decoder have the same structure

Another use of the autoencoder is to encode data for a supervised learning network. For instance, we could use the encoded variables as inputs to a fully connected feed-forward network. Because the input dimension to the network has been reduced, this network might need less training data to learn than a model that uses the full N inputs. Later on, we will see that this can be a powerful technique when dealing with high-dimensional input data.

> **Fully Connected Autoencoder**
> A feed-forward network in the form of those discussed in Chap. 5 can be used to fit the identity: $\mathbf{f}(\mathbf{x}) = \mathbf{x}$. If the network structure is such that one of the hidden layers has fewer units than there are independent variables in $\mathbf{x}$, we can consider the values of the hidden units as a low-dimensional representation of the inputs that we call codes. A neural network that produces these codes is called an autoencoder.
>
> - When a feed-forward neural network is used as an autoencoder, it is called a fully connected autoencoder.
> - Autoencoders typically have an hourglass structure where the middle hidden layer of the network is narrower than the input or output layer.
> - The codes discovered in the training of the autoencoder can be used to compress the data, understand hidden structure in the data, and to be inputs to other models.

8.2 Compressing Leaf Spectra with an Autoencoder

In Sect. 4.5, we used clustering techniques to understand the different transmittance and reflectance spectra from 276 different sample leaves. Each leaf sample had the reflectance and transmittance of the leaf measured for light at 2051 wavelengths from 400 to 2450 nm in increments of 1 nm. We will now use an autoencoder to reduce this data. The data set we are dealing with is of size $276 \times 2 \times 2051 = 1\,132\,152$. The autoencoder network we build has an encoder that has

1. An input layer of 2051 nodes,
2. A hidden layer with 64 nodes,
3. A hidden layer with 32 nodes, and
4. A hidden layer with ℓ nodes.

The value of ℓ will vary and the size of the encoded data, i.e., the latent variables/codes, will be of size ℓ. The decoder starts from the ℓ encoded values and then has

1. A hidden layer with 32 nodes,
2. A hidden layer with 64 nodes, and
3. An output layer with 2051 nodes.

In the decoder, there are 136 051 parameters to fit (weights and biases). All of the hidden units use the ReLU (rectified linear unit) activation function. Therefore, to store the entire data set in compressed format we will need $\ell \times 276 \times 2$ pieces of information for the reflectance and transmittance spectra, and we will also have to store the parameters in the decoder. We built two networks with $\ell = 2, 4$. The required storage of this information is then 12.1% and 12.2% of the full data set for $\ell = 2$ and $\ell = 4$, respectively. The percentage did not change very much because most of the storage is for the parameters in the decoder.

We trained the autoencoder over 500 epochs with a batch size of 32 for both cases and used the entire data set for training. The trained model produced a mean-absolute error for the data set of 0.0104 and 0.00557 for the $\ell = 2$ and $\ell = 4$ models. If we look at some representative cases from the training data, see Fig. 8.3, we observe that the autoencoded values are capturing the reflectance and transmittance spectra for the leaves well. In the figure, we can see that the autoencoder with a latent space of size $\ell = 2$ does have trouble getting the correct level of some of the features, but overall it captures the shape of the spectra. The autoencoder with latent space of size $\ell = 4$ does appear to have an overall better accuracy, as we would expect from the mean-absolute error numbers.

From the autoencoder, we can try to interpret what the latent space variables mean. In the case of a latent space of size two, we can make a scatter plot of all of the spectra with the latent space variables denoted as z_1 and z_2. This scatter plot is shown in Fig. 8.4a where the points have the color associated with each spectra, as discussed in Sect. 4.5. We can see there are two clusters that appear in the scatter plot, with most of the points centered in the top left of the figure. If we use the decoder, we can explore how changing a latent variable changes the resulting spectrum. For instance, in Fig. 8.4b we see that increasing z_1 increases the overall spectrum level when we keep z_2 fixed. However, we can also see that the effect is nonlinear: The shape of each resulting spectra is different and is not just a scaling as we would see in a singular value decomposition of the data. Figure 8.4c, d also display this nonlinear behavior when the values of z_2 are varied with z_1 fixed or when both are varied together. This nonlinear behavior is what allows an autoencoder to have such a small latent space that can still represent the data well.

8.3   Convolutional Autoencoders

The concept of an autoencoder can be extended to be a convolutional neural network (CNN). CNNs were discussed in Chap. 6. In this case, we can take a set of images and find a reduced representation of the images. Using a CNN makes sense because there will be geometric structure in images that are naturally suited to the structure of the convolutions learned by the network.

The convolution autoencoder, like a fully connected autoencoder, has the first half of the network act as an encoder, and the second half as a decoder. In this network, as depicted in Fig. 8.5, we use convolutions to compress and expand the data. In the encoder half we apply convolutions, as in Chap. 6, to create extra channels from the original data. To shrink the size of the data, rather than use pooling, we use strides in the convolution. That is, we do not center the convolution

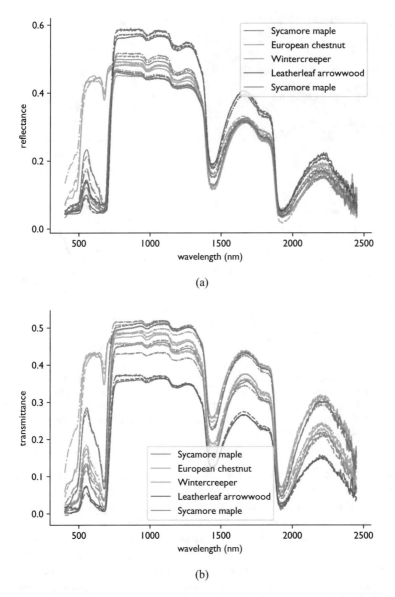

Fig. 8.3 Comparison between the autoencoder results using latent spaces of size 2 (dash-dot lines), size 4 (dashed lines), and the original data (solid lines) for the leaf reflectance and transmittance data. (**a**) Reflectance. (**b**) Transmittance

at every pixel rather we skip some. The stride tells the convolution how many pixels to move in each direction when applying the convolution. Therefore, a stride of $(1,1)$ does not skip any pixels, but a stride of $(2,2)$ will skip every other row and column in the image. This is shown in Fig. 8.6 where a 9×9 input is convolved with a 3×3

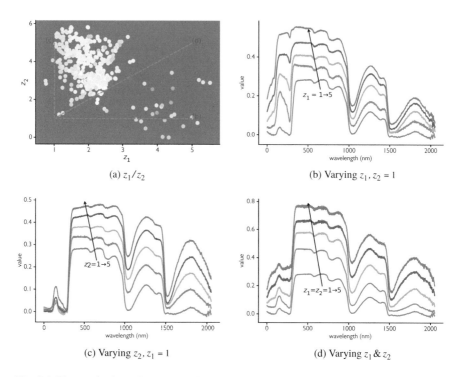

Fig. 8.4 The results from the autoencoder with latent space of size 2: (**a**) all the spectra (both reflectance and transmitted) on a scatter plot based on the values of the two latent space variables. The color of each point corresponds with the perceived color and the lines denote the path for the other three plots. (**b–d**) Plots of resulting spectra from feeding into the decoder values in unit increments along the lines shown in part (**a**)

kernel and stride (2,2) with zero padding of the input. Every input pixel contributes to the output, but every other row/column is skipped when moving the kernel over the input. This results in a reduced size for the output, and, as a result, a reduction of the data. The encoder combines several strided convolutional layers to compress the image into a set of code images, or latent images, in the middle of the network.

To expand the data from the code images to the original image, we apply convolutions in reverse through transposed convolutions. Transposed convolutions act in the opposite way of a convolution: A convolution takes an image and computes a linear combination of the pixel values using a kernel to get a new value. The transposed convolution takes an image and distributes those values to a set of new pixels with weights defined by the kernel. In this sense, the transposed convolution works backwards to the standard convolution. Furthermore, we can use a stride in the transposed convolution to generate a larger image in the output. In a strided, transposed convolution we skip some of the pixels in the output to center the convolution around. By applying several layers of transposed convolutions, we expand the code images back to the original size.

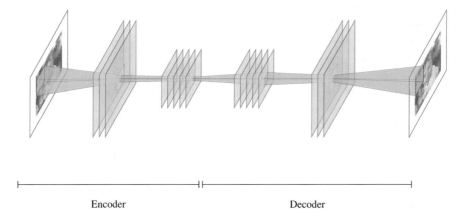

Fig. 8.5 Schematic of an autoencoder with multiple hidden layers. This particular network is symmetric about the encoded variables in that the encoder and decoder have the same structure

As with the fully connected autoencoder, if we can train a network that can reproduce the original data with acceptable accuracy, then we can store the coded images and use the decoder to reproduce the original data as needed. This could be useful in data transmission where rather than sending full images across a network, we send the smaller coded images and the receiver uses the decoder to reproduce the full image. Additionally, as we will see later that it is possible to use the codes as input to another neural network for a supervised learning problem. Using the codes as input reduces the input dimensionality to the problem and may make the network easier to train.

Convolutional Autoencoder

Convolutional autoencoders use convolutions to compress data, typically images, into a set of codes in a similar manner to the fully connected autoencoder.

- The encoder uses strided convolutions that only center the convolution at a fraction of the original inputs to reduce the data.
- Typically, while the size of the data is being reduced, extra channels are added.
- The decoder takes the codes and performs strided, transposed convolutions to project the reduced data contained in the codes back to the original size of the data.

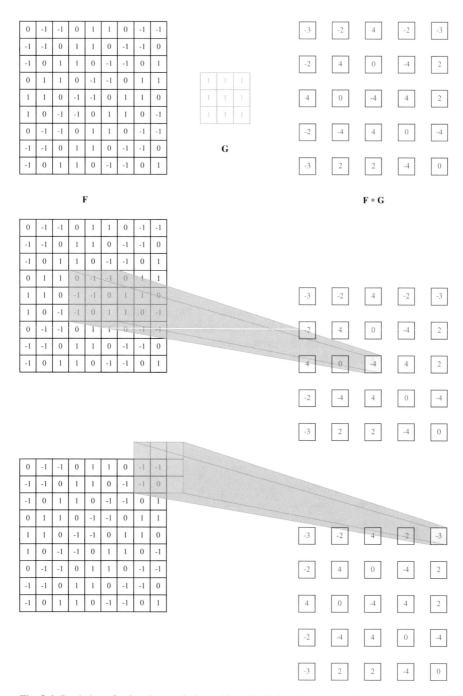

Fig. 8.6 Depiction of a 3×3 convolution with stride (2,2) using zero padding. The output is a 5×5 matrix whereas the input was 9×9

8.3.1 *Fashion MNIST Demonstration of Convolutional Autoencoders*

To demonstrate how convolutional autoencoders work, we turn to the Fashion MNIST data set introduced in Chap. 4. Each image is a 28×28 grayscale image of a different item of clothing, footwear, or accessory. The convolutional autoencoder we train has the following structure. The encoder has

1. An input layer of shape 28×28,
2. A convolutional layer with kernel size of 3×3, a stride of 2 in each direction, and 32 channels. This outputs 32 images of size 14×14;
3. A convolutional layer with kernel size of 3×3, a stride of 2 in each direction, and 16 channels. This outputs 16 images of size 7×7;
4. A convolutional layer with kernel size of 3×3, a stride of 1 in each direction, and 8 channels. This outputs 8 images of size 7×7;
5. A convolutional layer with kernel size of 3×3, a stride of 1 in each direction, and 4 channels. This outputs 4 images of size 7×7.

The result of this final layer is the encoded data. The size of this data is $7 \times 7 \times 4 = 196$; this is one-quarter the size of the original 28×28 image. For the decoder, we have

1. A transposed convolutional layer with kernel size of 3×3, a stride of 1 in each direction, and 8 channels. This outputs 8 images of size 7×7;
2. A transposed convolutional layer with kernel size of 3×3, a stride of 1 in each direction, and 16 channels. This outputs 16 images of size 7×7;
3. A transposed convolutional layer with kernel size of 3×3, a stride of 2 in each direction, and 32 channels. This outputs 32 images of size 14×14;
4. A transposed convolutional layer with kernel size of 3×3, a stride of 2 in each direction, and 1 channel. This outputs a single image of size 28×28.

The result of the decoder is the same size as the original image. Each layer uses ReLU (rectified linear unit) activation functions in the convolutions. The total number of parameters in the decoder is 5 343, or about the size of 7 of the original 28×28 images. Therefore, if we are storing a large number of these images, the cost to store the decoder is negligible.

In Fig. 8.7 we show the output of each of the layers of the autoencoder for two different cases. This figure details how the amount of data shrinks in the middle of the autoencoder. The encoding layers compute many different views of the input, but at successively smaller sizes. The code images in the middle of the network are much smaller than the original image, though there are four of them. In the decoder, we can see the effect of the stride in the transposed convolutional layers. The stride of 2 in each direction adds whitespace around certain pixels, when expanding the image. In the last layer, we apply a particular kernel to each of the 32 channels to get the single output.

Fig. 8.7 Application of the convolutional autoencoder to two example items from the fashion MNIST database. The autoencoder layers are arranged vertically: The original image is on top, followed by the 4 encoder layer outputs, and the 4 decoder layer outputs. The small block of 4 images are the codes, or latent images, for the autoencoder

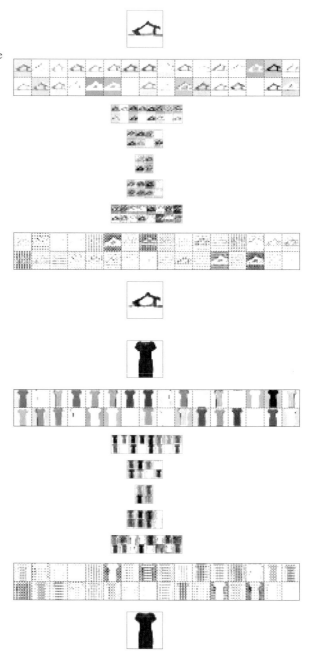

We should step back and think about what this model is doing. For the encoder, it is learning a set of convolutions to apply to the original image to create a smaller set of images that contain all of the information in the original image. This is possible

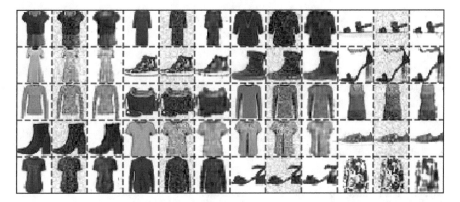

Fig. 8.8 Demonstration of the noise removal properties of the autoencoder. For each original item (left image), we added random noise (middle image) and passed the noisy image through the autoencoder to get the third image. We can see that the autoencoder successfully removes the noise around the item and improves the image

because the original images have structure to them: The images are not just random pixels. Then the decoder learns a set of nonlinear transformations to turn the codes into the original image. Both of these tasks, encoding and decoding, are being learned at the same time because our goal is to have the result of the decoder be as close to the original image as possible.

8.3.2 Noise Reduction with Convolutional Autoencoders

Beyond compressing the data, autoencoders can remove noise from an image. In the fashion MNIST example, we trained the network with images without noise. If we pass a noisy image through the network, the resulting image can have less noise. This is because the random noise in the image is removed by the encoding process.

We demonstrate this effect by adding Gaussian noise with mean zero and standard deviation equal to 10% of the total pixel range to example images from the fashion MNIST data set. We then pass the noisy image through the autoencoder. As we see in Fig. 8.8, the result from the autoencoder effectively removes noise from the image, especially the noise that corrupts the whitespace around the item. This is a beneficial side effect of the autoencoder being trained to find only the important features in the image for reconstruction. We could improve this noise filtering property by training the autoencoder with noisy inputs but with target images that do not have noise. This would be useful, if one expects to use the autoencoder with noisy data.

8.4   Case Study: Reducing the Data from a Physics Simulation with Autoencoders

Autoencoders have found use in reducing the size of scientific simulation data in order to reduce the dimensionality of simulation data to then use as inputs to other machine learning models [1, 2]. We will demonstrate how this works using the simulations of laser irradiated targets. The target, called a hohlraum, is made of gold and has the shape given in Fig. 8.9. This shape is loosely based on the Apollo hohlraum used in opacity experiments [3, 4].

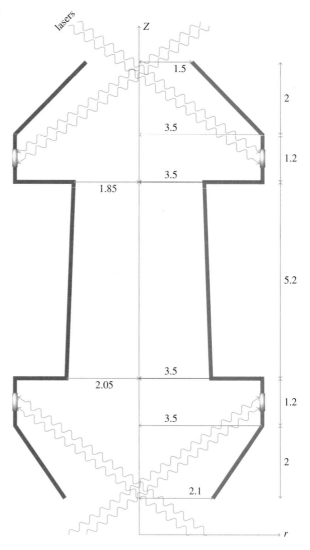

Fig. 8.9 The nominal Apollo hohlraum used in the simulations. This diagram shows the outline of the hohlraum target in cylindrical $(r - z)$ coordinates with dimensions given in millimeters. The laser beams strike the target as depicted here; in simulations the laser is represented as a radiation source in the areas indicated by a rectangles in the flat part of the outer chamber

In an experiment a laser irradiates the target, causing the material to heat up and radiate as a blackbody in the X-ray regime [5]. The rapid heating of the gold causes it to vaporize into a plasma that expands from the wall. In this case, study we consider only the movement of the radiation from the heated area and simulate how it moves about the evacuated cavity of the hohlraum. We use an approximate method known as flux-limited diffusion [6] to simulate the propagation of photons as a demonstration (typically a more accurate method such as implicit Monte Carlo [7], discrete ordinates [6], or spherical harmonics [8] would be used). To simulate the laser heating, we place a radiation source in the hohlraum walls in the areas indicated by rectangles in Fig. 8.9. We watch the radiation fill the inner cavity of the hohlraum over time. For the experiments, we are interested in the time history of the radiation energy in the cavity that we quantify as a radiation temperature, i.e., the equivalent blackbody temperature corresponding to the radiation energy density at a particular location. The simulations use a 2-D cylindrical geometry with a regular grid of 74 and 120 zones in the r and z directions, respectively; each simulation runs for 500 time steps.

For our study, we will be perturbing the dimensions of the hohlraum and the length of the laser pulse. We then simulate the laser irradiation and response of the hohlraum for a fixed time. In particular we define 4 parameters that we vary:

1. A scale parameter that multiplies each dimension of the hohlraum,
2. An asymmetry parameter that multiplies the opening radius at the top of the hohlraum and divides the nominal radius at the bottom,
3. An interior scale parameter that multiplies the nominal radii of the edges of the interior of the cavity (1.85 and 2.05 mm in Fig. 8.9), and
4. A laser pulse scale parameter that multiplies the length of the laser pulse; the laser power is divided by this parameter so that the total energy deposited is constant.

Between these four parameters, it is possible to vary the overall size of the hohlraum, the relative asymmetry of the openings, and the width of the interior chamber at the same time. If all of the parameters are 1, then we are considering the nominal hohlraum. We vary each of the parameters from 0.8 to 1.2 to, for instance, shrink the hohlraum by a factor of 0.8 in all dimensions and to enlarge it up to a factor of 1.2. We use Latin hypercube sampling to select 64 different values for the four parameters [9]. The values of the four parameters used in these 64 simulations are shown in Fig. 8.10.

The time history for a typical simulation is shown in Fig. 8.11. Here we see that at early time the areas heated by the laser quickly reach a high temperature. This creates radiation that begins to fill the cavity. By 1.79 ns, the cavity has a large amount of radiation in it, and as the laser turns off the light eventually leaks out of the hohlraum. This is typical of the simulations for different parameter settings though the timings and the maximum temperature will be different as we scale the geometry and the laser pulse length. Additionally, we have seen that in some cases the radiation can leak through the wall of the cavity and enter the area between the outer lobes of the target.

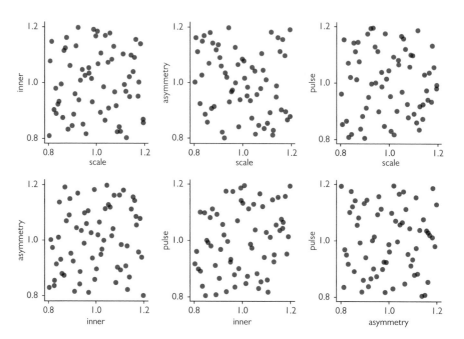

Fig. 8.10 Scatter plots of the 64 values of the four parameters (overall scale, asymmetry, interior scale, and laser pulse scale) defining the hohlraum geometry and the laser pulse for the simulation data set

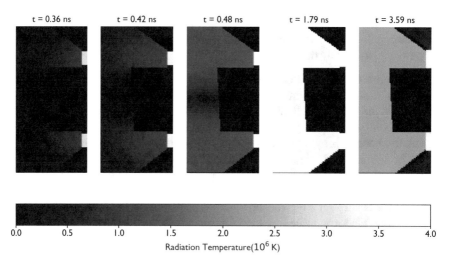

Fig. 8.11 Depiction of the radiation temperature in the cavity at several times showing the cavity filling with radiation and then cooling after the laser turns off. The horizontal axis is the radial coordinate and the vertical axis is z

From the 64 simulations and 500 time steps for each, we have $500 \times 64 = 32\,000$ 2-D arrays for the radiation temperature in the hohlraum. Looking at Fig. 8.11, we can see that there is an obvious structure to the outputs from the simulation. We would like to understand if there is a reduced representation of the data that can be found with a convolutional autoencoder. If we are able to find a suitable latent space representation of the data, we will use these latent variables as the inputs to an LSTM network to predict the time history of the radiation temperature in a given simulation.

8.4.1   Convolutional Autoencoder Set Up and Training

The data that we use to train the model consists of 80% of the $32\,000$ time steps from the simulation outputs. We also pad the images to be of size 128×80. This is done so that we can repeatedly reduce the image in each dimension by a factor of 2. Furthermore, we compute the maximum value of the radiation temperature in each image and store this value. We then normalize each image by dividing by the maximum radiation temperature. This makes each image in our set have each pixel have a value between 0 and 1.

The autoencoder network we train has a similar structure to that used for the Fashion MNIST data. The encoder portion of the network consists of 3 convolutional layers that compress the original image from $128 \times 80 = 10\,240$ to 8 images of size 16×10 for a total code size of $8 \times 16 \times 10 = 1\,280$ or about 12.5% of the original size. The decoder portion of the network also has three layers to increase from the codes back to the original image size.

In detail, the encoder portion of the network contains

1. An input layer of shape 128×80,
2. A convolutional layer with kernel size of 3×3, a stride of 2 in each direction, and 32 channels. This outputs 32 images of size 64×40;
3. A convolutional layer with kernel size of 3×3, a stride of 2 in each direction, and 16 channels. This outputs 16 images of size 32×20;
4. A convolutional layer with kernel size of 3×3, a stride of 2 in each direction, and 8 channels. This outputs 8 images of size 16×10.

The result of this final layer is the encoded data. The size of this data is $16 \times 10 \times 8 = 1280$; this is about 14% the size of the original 120×74 image. For the decoder, we have

1. A transposed convolutional layer with kernel size of 3×3, a stride of 2 in each direction, and 16 channels. This outputs 16 images of size 32×20;
2. A transposed convolutional layer with kernel size of 3×3, a stride of 2 in each direction, and 32 channels. This outputs 32 images of size 64×40;
3. A transposed convolutional layer with kernel size of 3×3, a stride of 2 in each direction, and 1 channel. This outputs an image of size 128×80.

All of the layers use the ReLU activation function.

After training for 100 epochs using 512 cases per batch using the Adam optimizer. For both the training and test data, the mean-absolute error was approximately 0.0108 or roughly 1% when comparing the images reconstructed from the codes. In Fig. 8.12, we show a representative case from the test data. At the top is the original image from the simulation data. Below that we show the 32 images that are the output of the first layer of the encoder, each one-half the size in each of the horizontal and vertical dimensions. The number of outputs images from each of the subsequent two layers is reduced at the same time the size of the image is reduced. Note in Fig. 8.12 that some of the images in the intermediate layer highlight the edges of the hohlraum while others have convolutions that measure the values in the interior of the cavity. Additionally, the vertical and horizontal edges of the hohlraum are detected by different filters in the encoder network. In many respects, this is a challenging problem for the autoencoder because the narrow part of the hohlraum is not aligned with the horizontal or vertical direction, and the angle it makes is different when the geometry changes. Despite this, the autoencoder appears to be effective in finding a set of codes or latent variables that can faithfully describe the full set of images from the radiation temperature as a function of time from the simulations.

Removing Noise with the Autoencoder

As we saw previously the data reduction properties of the autoencoder can be used to remove noise from data. We demonstrate this in the hohlraum simulation data by taking the original values for the radiation temperature and adding random Gaussian noise with standard deviation of 0.1 (or 10% of the maximum value in the normalized images) and passing the data with noise through the autoencoder. The results for this noise reduction for 20 different images are shown in Fig. 8.13. The resulting images remove a large portion of the noise from the resulting image with minimal distortion of the geometry of the hohlraum. These results indicate that autoencoders might be able to improve the results from numerical methods such as particle Monte Carlo methods that have spurious noise in the simulation results.

8.4.2 LSTM with Inputs from the Autoencoder

One benefit to the codes from the autoencoder is that they contain the information from the original data. Therefore, rather than using the original data as input to a prediction model, we can use the codes as a reduced image. Given that the size of the input data is reduced, it is likely that fewer training examples and/or simpler models can be used too. In our case, we would like to use the results from a few initial time steps to predict the later time behavior of the radiation temperature using a LSTM model (see Chap. 7). Without using the autoencoder, we would need to attempt this

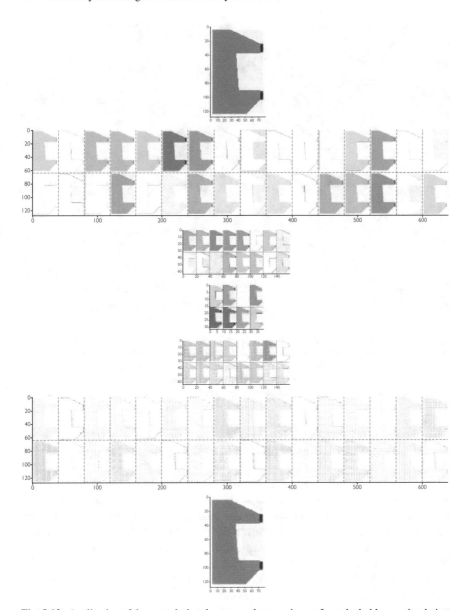

Fig. 8.12 Application of the convolutional autoencoder to an image from the hohlraum simulation data. The autoencoder layers are arranged vertically: The original output is on top, followed by the 3 encoder layer outputs, and the 3 decoder layer outputs. The small block of 4 images are the codes, or latent variables, for the autoencoder. The images are shown with a color scale where black represents a value of 1 and white is a value of 0

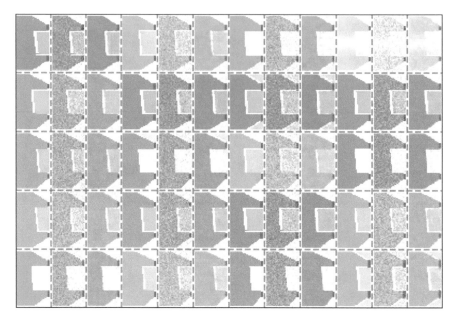

Fig. 8.13 Demonstration of the noise removal properties of the autoencoder for the hohlraum data. For each item we added random noise (middle image) and passed the noisy image through the autoencoder to get the third image. We can see that the autoencoder successfully removes the noise around the item and improves the image when compared to the original on the left. The images are shown with a color scale where black represents a value of 1 and white is a value of 0

using the full data (the 120×74 images) from each time step or we could try to use domain expertise such as knowledge of the physics of the simulation to reduce the data [10].

With the autoencoder we trained above, we can train an LSTM model to take a sequence of the codes, each of size $8 \times 16 \times 10$ and then predict future values for the codes. Then using the decoder, we can feed the output of this model into the decoder to get the radiation temperature at the full 120×74 size. The LSTM model will also have to predict the maximum value for the radiation temperature so that we can undo the normalization.

Our model is designed to take the codes corresponding to the simulation data and the maximum value of the radiation temperature at ten time points, each spaced 20 time steps apart. It will predict the codes and the maximum values of the radiation temperature at ten subsequent times separated 20 time steps apart. The total number of inputs to the model is $(16 \times 10 \times 8 + 1) \times 10 = 12\,810$ and an equivalent number of outputs. The model consists of two connected LSTM units each with 16 states. The second LSTM unit is connected to two dense layers each with 10 neurons using the ReLU activation function. The output layer has $12\,810$ units to give codes and normalization constants at ten time levels. From this data, we can apply the decoder to get the full image.

We select at random 10^5 starting times and use 80% of these as training values. We train the model for 100 epochs with the Adam optimizer and obtain a mean-absolute error of 0.0176 for the training data and 0.0177 for the test data. Predictions from two different time series for the predicted and actual radiation temperatures are shown in Fig. 8.14. In the figure, we can see that the model is able to predict the behavior for different regimes of evolution. In the top case, the laser is still on and the temperature is not decreasing in the early frames, and the bottom case depicts after the laser has turned off and radiation is leaving the hohlraum.

One feature we notice in the results from the model is that the edge of the hohlraum in the narrow part of the cavity has a noticeable amount of error. This is especially noticeable in the first frame of the lower example in Fig. 8.14. Nevertheless, it appears on the scale of the figure that away from the boundary there is agreement between the predicted and actual values.

Figure 8.15 shows the values of the radiation temperature at the center of the narrow part of the cavity for four different test cases. The behavior of the radiation temperature has different shapes depending on the starting time, geometry, and laser pulse length. We observe that the model is able to predict the shape of the behavior, but there is some inaccuracy in the numeric values. These results demonstrate the model would be useful for analyzing hohlraum design and predicting the radiation temperature based on the geometry and laser pulse parameters. Once a design was selected, a full simulation could be performed to get more accurate numerical values for the temperature.

Notes and Further Reading

Unsupervised learning with neural networks is a broader topic than just autoencoders. Indeed in there are variations of autoencoders, called variational autoencoders, that attempt to find not just latent variables but the distributions that govern these variables. The background required to understand these tools would take us too far afield for our study, but interested readers can see the tutorial by Doersch [11]. The benefit of a variational autoencoder is that one can sample from the distribution of latent variables to produce new cases. In the case of Fashion MNIST, one could generate new items of clothing.

Other types of unsupervised learning techniques include Boltzmann machines and Hopfield networks. These approaches to understanding structure in data were introduced many years ago and have not been as widely adapted as autoencoders. The restricted Boltzmann machine has been used to create generative networks that given a set of data can learn the probability distribution of its inputs. Restricted Boltzmann machines can also be used to initialize autoencoders and speed up training [12].

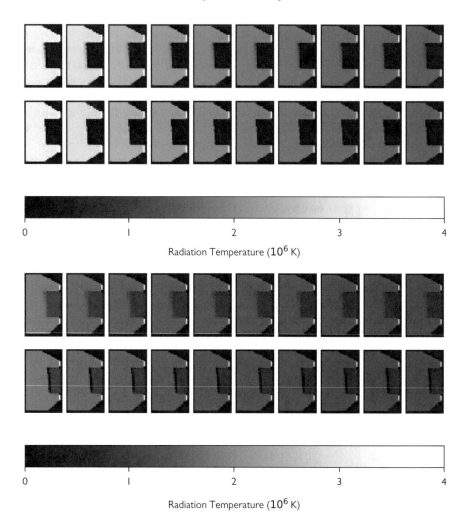

Fig. 8.14 Comparison of the actual and predicted values for the evolution over 10 time levels of the radiation temperature in two cases. In each case, the top row contains the values predicted by the LSTM model and passed through the decoder, and the lower row is the actual values of the codes passed through the decoder model

Problems

8.1 The latent variables or codes in an autoencoder attempt to capture something about the true number of degrees of freedom in a data set. Produce a data set of the value of $\sin(at + b)$ at points $t = 0, 0.05, 0.2, \ldots, 0.95$ and for values of a and b selected at random between -1 and 1. Build a fully connected autoencoder that takes as input the value of the $\sin(at + b)$ at the twenty time points and tries to reproduce

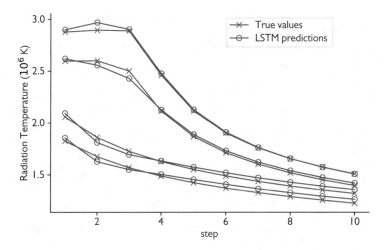

Fig. 8.15 Values for the predicted at actual values for the radiation temperature in the middle of the narrow part of the cavity for ten time values. There are four different time series shown

the original signal. Theoretically, the autoencoder should be able to have a middle layer that only has two nodes because there are only 2 degrees of freedom in the data you created. How close to this theoretical limit can you get?

8.2 This is a variation on the previous problem, but using convolution. Produce data from the function $\sin(ar + b)/r$ where $r = \sqrt{x^2 + y^2}$ for values of x and y given by $-0.95, -0.9, \ldots, 1$ and values of a and b selected at random between -1 and 1. Each case will be an image of size 40×40. Apply a convolutional autoencoder and see how well you can compress the data.

References

1. Kelli D Humbird, J Luc Peterson, and Ryan G McClarren. Parameter inference with deep jointly informed neural networks. *Statistical Analysis and Data Mining: The ASA Data Science Journal*, 12(6):496–504, 2019.
2. John E Herr, Kevin Koh, Kun Yao, and John Parkhill. Compressing physics with an autoencoder: Creating an atomic species representation to improve machine learning models in the chemical sciences. *The Journal of chemical physics*, 151(8):084103, 2019.
3. Tana Cardenas, Derek William Schmidt, Evan S Dodd, Theodore Sonne Perry, Deanna Capelli, T Quintana, John A Oertel, Dominic Peterson, Emilio Giraldez, and Robert F Heeter. Design and fabrication of opacity targets for the national ignition facility. *Fusion Science and Technology*, 73(3):458–466, 2018.
4. Evan S Dodd, Barbara Gloria DeVolder, ME Martin, Natalia Sergeevna Krasheninnikova, Ian Lee Tregillis, Theodore Sonne Perry, RF Heeter, YP Opachich, AS Moore, John L Kline, et al. Hohlraum modeling for opacity experiments on the national ignition facility. *Physics of Plasmas*, 25(6):063301, 2018.

5. R Paul Drake. *High-energy-density physics: foundation of inertial fusion and experimental astrophysics*. Springer, 2018.
6. Thomas A Brunner. Forms of approximate radiation transport. *Sandia report: SAND2002-1778*, 2002.
7. Allan B Wollaber. Four decades of implicit Monte Carlo. *Journal of Computational and Theoretical Transport*, 45(1-2):1–70, 2016.
8. Ryan G McClarren and Cory D Hauck. Robust and accurate filtered spherical harmonics expansions for radiative transfer. *Journal of Computational Physics*, 229(16):5597–5614, 2010.
9. Ryan G McClarren. *Uncertainty Quantification and Predictive Computational Science*. Springer, 2018.
10. Ryan G McClarren, D Ryu, R Paul Drake, Michael Grosskopf, Derek Bingham, Chuan-Chih Chou, Bruce Fryxell, Bart van der Holst, James Paul Holloway, Carolyn C Kuranz, Bani Mallick, Erica Rutter, and Ben R Torralva. A physics informed emulator for laser-driven radiating shock simulations. *Reliability Engineering and System Safety*, 96(9):1194–1207, September 2011.
11. Carl Doersch. Tutorial on variational autoencoders. *arXiv preprint arXiv:1606.05908*, 2016.
12. G E Hinton. Reducing the Dimensionality of Data with Neural Networks. *Science*, 313(5786):504–507, July 2006.

Chapter 9
Reinforcement Learning with Policy Gradients

You should be studying your arts, instead of studying me!
That's how you lost your first job, punk!
Now get in line...

—Ghostface Killah *from the song "Nutmeg" on the album*
Supreme Clientele

Abstract In this chapter we provide an introduction to a different type of learning problem called reinforcement learning. In reinforcement learning problems we do not know the correct value for the dependent variable, as we would for a supervised learning problem, but we do have an objective function called a reward. We want our machine learning model output to maximize the reward given its inputs. Additionally, the model might need to produce a series of correct answers to get an even bigger reward. We motivate this discussion using a cart-mounted pendulum where to get the maximum reward the pendulum must be successively swung back and forth to get a full rotation. Our method for reinforcement learning is known as policy gradients. In policy gradients we transform the reinforcement learning problem into a supervised learning problem where the dependent variables are weighted by the reward during training. This trains the model to produce outputs that maximize the reward and not produce outputs with a small reward. We conclude the chapter with a case study, where we use policy gradients to control the cooling of glass in an industrial process.

Keywords Reinforcement learning · Policy gradients · Reward functions · Exploring solution space · System control

9.1 Reinforcement Learning to Swing the Cart-Mounted Pendulum

We have previously discussed supervised and unsupervised learning problems. In this chapter we deal with a variant on supervised learning called reinforcement learning. In reinforcement learning problems we have an objective function that

© Springer Nature Switzerland AG 2021
R. G. McClarren, *Machine Learning for Engineers*,
https://doi.org/10.1007/978-3-030-70388-2_9

we would like to maximize by making a series of decisions using a model, but we do not have any information on how to maximize that objective. We train a machine learning model to learn about the objective function and what actions lead to higher values of the objective function.

In reinforcement learning we often think of the function that we are trying to maximize as a reward function. In this terminology we want to learn what actions to take by the model to maximize the reward. The training of such a model will be different from supervised learning because we will need to take the reward function and make different actions to determine how the actions lead to rewards. In our study we use a type of reinforcement learning called policy gradients: that is we devise a decision-making structure (a policy) and use gradient descent to find the policy that maximizes our reward. These types of algorithms have been shown to be successful in devising strategies for getting high scores, even superhuman scores, in certain video games [1].

We are going to motivate the study of reinforcement learning using the cart-mounted pendulum. Previously, we discussed the problem of the evolution of a pendulum mounted on a cart in Sect. 7.3. In this problem the swinging of the mass causes the cart to move. In this case we will allow the cart to apply a force via its wheels to accelerate the cart to the left or right. This changes the equations of motion of the system to be [2]

$$\ddot{x} = \frac{\hat{f} + m \sin\theta \left(\ell\dot{\theta}^2 + g\cos\theta\right)}{M + m\sin^2\theta}, \tag{9.1}$$

$$\ddot{\theta} = \frac{-\hat{f}\cos\theta - m\ell\dot{\theta}^2\cos\theta\sin\theta - (M+m)g\sin\theta}{\ell(M + m\sin^2\theta)}. \tag{9.2}$$

In these equations, $\hat{f}$ is the force applied which we allow to be one of the three values $-10, 0$, or 10 N. Additionally, we fix the mass and the cart to each be 1 kg ($m = M = 1$ kg), the length of the pendulum arm to be 1 m, and the acceleration due to gravity to be $g = 9.81$ m/s^2.

The goal for our reinforcement learning problem is to learn a strategy for the cart to begin with the pendulum at the bottom of its swing ($\theta = 0$) and to swing the pendulum to the vertical position of $\theta = \pm\pi$ and balance it there for as long as possible. This is a more difficult problem than it might sound because the cart must accelerate in one direction and then rapidly accelerate in the other direction to swing the pendulum to be completely vertical. In other words, to accomplish its task our machine learning model must understand that there are times when it wants the pendulum to go to the bottom in order to get the pendulum to a high height later.

The process of swinging the mass to be upright is illustrated in Fig. 9.1. When the mass is at rest, the cart accelerates in one direction (right in the figure), which causes the mass to swing up behind it (see panel 2 in the figure). Then when the mass reaches its apex in panel 3, the cart accelerates in the other direction, giving the mass more speed as it falls (panel 4). This gives it enough momentum to reach an upright position (panels 5 and 6).

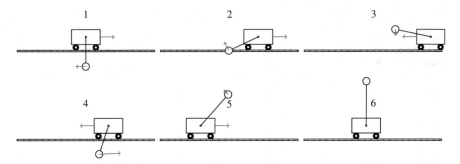

Fig. 9.1 Illustration of the strategy to swing the mass of the cart-mounted pendulum from rest to the maximum height. The numbers in each figure indicate a step in the process; the arrow on the cart indicates the force applied by the cart, and the arrow from the mass indicates direction it is moving and its speed

> **Reinforcement Learning**
> Reinforcement learning aims to train a model to maximize a reward function. It does not need to know the correct answer for training, unlike supervised learning. In reinforcement learning with deep neural networks, we design a training procedure that will find how to optimize the weights and biases in a neural network to maximize the reward.

9.1.1   Reward Function

To fully state the problem, we need to define our objective function or reward. This will encode the goal of our reinforcement learning problem that we wish to solve. Given the goal that we have defined, for the reward at a given time, we evaluate the following function:

$$\rho(\theta) = \frac{1}{(\pi - |\theta|)^4 + \epsilon} + b, \tag{9.3}$$

where

$$b = \begin{cases} 20(y + 1) & y \geq -\frac{1}{2} \\ 0 & y < -\frac{1}{2}, \end{cases} \tag{9.4}$$

and the height of the pendulum, y, is given by

$$y = \cos(\pi - \theta). \tag{9.5}$$

Fig. 9.2 Reward function for the cart-mounted pendulum problem. The function is proportional to $(\pi - |\theta|)^{-4}$ with an added bonus when $|\theta|$ crosses $2\pi/3$

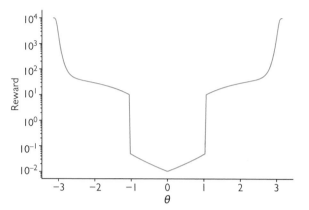

The reward function we have specified has two terms and is plotted in Fig. 9.2. The first term rewards the system for getting the angle between the pendulum and the cart to be close to $\pm\pi$. The form of this function makes the reward increase nonlinearly as the goal is approached. Moreover, once the pendulum gets above a certain height, y, there is a "bonus" reward, b, that kicks in. When the height of the pendulum mass is greater than -0.5 (or when $|\theta| \geq 2\pi/3$), there is a bonus applied that increases the magnitude of the reward that is linear in y. The purpose of the bonus is to indicate that getting the pendulum to high level will lead to higher rewards. This will be discussed in more detail later. For now we can think of it as analogous to a bonus in a video game when a certain threshold is crossed. If the cart system were a video game, we could imagine the screen flashing the words "super bonus" when the height of the pendulum gets high enough.

Reward Functions
The reward function is where we encode the desired behavior of the system numerically. It can be important that reward function is designed so that intermediate rewards exist that can be found via random sampling.

9.1.2 Controlling the Cart Pendulum System

We use the same differential equation integrator from Sect. 7.3 to update the state of the system given the current position of the cart, x, and its time derivative, $\dot{x}$, the value of θ and its time derivative, $\dot{\theta}$, and the force applied. The integrator updates the state to a time Δt forward in time $1/32 = 0.03125$ s, and the force applied over this time is constant. Before each update, our controller function will be given $(x, \dot{x}, \theta, \dot{\theta})$ as input, and it must output the force to be applied, either -10, 0, or 10 N.

The controller will have 200 steps of size Δt to try to maximize the reward. If we call one realization of these 200 steps a case, the total reward, R, for the case will be the sum of the rewards at the end of each step:

$$R = \sum_{i=1}^{200} \rho(\theta(i\,\Delta t)), \tag{9.6}$$

with $\theta(i\,\Delta t)$ the value of θ at time $i\,\Delta t$.

We have fully defined the physical system, the reward function, and how the controller interacts with the system. Nevertheless, we need to specify how the controller will make a decision.

9.2 Training a Neural Network with Policy Gradients

For the problem described above for the cart-mounted pendulum, or for any scenario where we want to take actions to maximize a reward function, we consider the problem of having a neural network to make decisions on what actions to take. The act of taking actions based on current conditions (i.e., the state of the system) is called a policy. For example, a policy could always apply a force of the opposite sign as the derivative of θ in the pendulum problem. In our case the neural network will embody the policy. It takes in inputs and chooses a decision to make. This network will output the probability that a given action should be taken by having the last layer of the network be a softmax function applied to a number of hidden units equal to the number of possible decisions. This is the policy function, $\mathbf{f}(\mathbf{x})$, where $\mathbf{x}$ is a vector of the same size as the number of inputs and $\mathbf{f} = (f_1, f_2, \ldots, f_N)$ is a vector of N probabilities containing the probability of choosing a particular action. For the cart-mounted pendulum problem, $\mathbf{x}$ is of length 4 containing $(x, \dot{x}, \theta, \dot{\theta})$. The policy function outputs 3 values: the probability of applying a force of -10 N, the probability of applying a force of 0 N, and the probability of applying a force of 10 N.

In the application of the policy function provided by the neural network, we evaluate $\mathbf{f}(\mathbf{x})$ to get the probabilities for each action. We then take the cumulative sum of the probabilities:

$$s_\ell = \sum_{n=1}^{\ell} f_n, \quad 1 \le \ell \le N. \tag{9.7}$$

Selecting a random number between 0 and 1 called ω, the action taken is the largest value of ℓ such that $\omega < s_\ell$. For instance, if $\mathbf{f} = [0.1, 0.3, 0.6]$, the first action would be selected for ω less than 0.1, the second action would be selected for $0.1 \le \omega < 0.4$, and the third action would be taken for $\omega \ge 0.4$.

Given that we have a network and a means to evaluate $\mathbf{f}(\mathbf{x})$ and then select an action, the question remains how to train this neural network. We want to train the network to maximize the reward for a case, but the function only outputs probabilities for making a decision. How can we adjust the weights and biases of the model to maximize the reward function? To do this, we will have to weigh the outputs of the policy function with the reward.

To see how to train the model, we consider a problem with a reward function $\rho(n)$, where n is the action taken by the controller and $f_n(\mathbf{x})$ is the probability of the policy choosing action n. The policy function will depend on the weights and biases of the neural network; we denote the weights and biases as $\mathbf{w}$. Given this scenario, the expected value for the reward, $E[\rho]$, is equal to the sum of the rewards for each choice n times the probability of making the choice. In our case, we have

$$E[\rho] = \sum_{n=1}^{N} \rho(n) f_n(\mathbf{x}). \tag{9.8}$$

This equation says that the reward we can expect is equal to the weighted sum of the potential rewards where the weight is given by the probability of selecting a given action. To apply stochastic gradient descent to increase the reward, we need to know what the gradient of the expected value of the reward is with respect to $\mathbf{w}$. Taking the derivative of the expectation with respect to a single weight, w, we get

$$\frac{\partial}{\partial w} E[\rho] = \frac{\partial}{\partial w} \sum_{n=1}^{N} \rho(n) f_n(\mathbf{x}) \tag{9.9}$$

$$= \sum_{n=1}^{N} \rho(n) \frac{\partial}{\partial w} f_n(\mathbf{x}).$$

Note that the derivative can be taken into the sum and that the reward function does not depend on the weight. Now we use an identity for derivatives that says

$$\frac{\partial g}{\partial x} = \frac{g(x)}{g(x)} \frac{\partial g}{\partial x} = g(x) \frac{\partial}{\partial x} \log g, \tag{9.10}$$

because the derivative of the natural logarithm is one divided by its argument. Using this identity in Eq. (9.9), we get

$$\frac{\partial}{\partial w} E[\rho] = \sum_{n=1}^{N} \rho(n) f_n(\mathbf{x}) \frac{\partial}{\partial w} \log f_n(\mathbf{x}) \tag{9.11}$$

$$= E\left[\rho(n) \frac{\partial}{\partial w} \log f_n(\mathbf{x})\right].$$

This final equation tells us that the derivative of the expected reward is equal to the expected value of the reward times the derivative of the policy function with respect to the weight. We can compute the gradient of the logarithm of the policy function with respect to its weight using back propagation, as previously discussed. This is why the method is called policy gradients: we use the gradient of the logarithm of the policy function to update the weights and biases in the neural network. When we use the policy function to control our problem, only a single action will be selected at each step, and we will know the reward of taking this action. Therefore, we will know what $\rho(n)$ is for the selected case.

This still does not complete the description of how we train the network. To see how this is done, we consider a single decision point. We evaluate $\mathbf{f}(\mathbf{x})$ and select an action, $\hat{n}$. After selecting the action, we will know what the value of the reward is. We then treat the "correct" example, as if we were doing supervised learning, as $\hat{n}$, and compute the cross-entropy loss (c.f. Eq. (2.25)). The loss function for policy gradients is

$$L = -\sum_{i=1}^{I} \sum_{n=1}^{N} \rho(n) \mathcal{I}(n, \hat{n}) \log \operatorname{softmax}(n, z^{(1)}, \ldots, z^{(N)}), \tag{9.12}$$

where $\mathcal{I}(n, \hat{n})$ is the indicator function that takes on the value of 0 when its two arguments are not identical and 1 when they are. This loss function has the same form as that in Eq. (9.11). Therefore, we weight the loss by the reward obtained by taking the action and train in the same manner as a standard classification problem. This is the crucial step in policy gradients: conversion of the reinforcement learning problem into a supervised learning problem where the cases are weighted by the reward of the selected action.

When the optimizer attempts to minimize the weighted loss function, it will try to increase the probability of selecting actions that lead to a good reward and decrease the probability of selecting the actions that lead to low rewards. This is exactly what we want our training to do: find a policy that maximizes the reward.

Policy Gradients
The policy gradients approach to reinforcement learning turns the reinforcement learning problem into a modified supervised learning problem. This is done by running the model to select an action, calling that action the "correct" dependent variable from a supervised learning perspective. When computing the loss function, for supervised learning we weight each case by the reward. This will make the trained model reproduce the actions that give large rewards and avoid the actions that give low rewards.

Choosing the Reward Weighting

The procedure outlined above, using the reward to weight the loss function with the "correct" selection being the choice made by the policy network used in the loss function, will cause the policy network to favor selections that maximize the reward in each step. However, as we pointed out for the cart-mounted pendulum, there are times when the controller should make a choice that decreases the reward in the short term in order to get a much larger reward later.

To address this problem, we can weight the loss function by the total reward, also called the cumulative reward, for a case as in the following:

1. run a case consisting of S steps,
2. compute the total reward, R, as the sum of the rewards for each step, and
3. add to the training data the inputs from each step $\mathbf{x}_s$, the selected actions, $\hat{n}_s$, and store the total reward R for $s = 1, \ldots, S$.

We repeat this process a given number of times and then run the optimization algorithm to update the neural network. Then we run the model for more cases and adjust the weights in the network. This goes on until the neural network reaches an acceptable level of reward.

It may take a large number of cases of operating the controller with the policy network to get an acceptable model. Given the way that reinforcement learning works in the early cases, the neural network will be making decisions that are basically random. It will take many cases of random decisions to get a large reward, if the problem is even moderately difficult. For instance, in the cart-mounted pendulum, we need the neural network to figure out, probably through luck, that it needs to accelerate in the opposite direction when the pendulum is reaching the top of a relatively low swing.

Cumulative Rewards

In policy gradients, and reinforcement learning in general, we want to optimize the total reward over all the actions taken by the model. Otherwise, if we maximize the reward in the next step, the model will not select actions that lead to delayed, but larger rewards. In policy gradients, this means we sum all of the rewards over a series of actions for the purpose of weighting the loss function.

9.3 Policy Gradient Training and Results for the Cart-Mounted Pendulum

We describe in detail the application of policy gradients to the cart-mounted pendulum problem. The neural network that we train is a fully connected feedforward network with

- an input layer with 4 inputs for the current state of the system: $\theta, \dot{\theta}, x, \dot{x}$,
- a hidden layer with 6 hidden units and the ReLU activation function,
- a hidden layer with 3 hidden units and the ReLU activation function, and
- a softmax output layer with 3 units for the three forces applied: $-10, 0$, and 10 N.

This network is trained using the policy gradients approach that we outlined above. The particular way that we train the network is as follows. A session of the "game" for the cart-mounted pendulum starts with the mass at rest at the bottom of its swing. The network is evaluated at this initial state, giving the probabilities for each of the 3 possible forces that can be applied, and one is selected based on sampling. Once the force, or move, is selected, the equations of motion for the pendulum are solved using the current conditions as initial conditions and with the force applied fixed at the sampled value for $1/32$ s. The result of this update gives the new conditions for the system, $\theta, \dot{\theta}, x, \dot{x}$, and the reward for the system at this new state. We repeat this procedure for 200 steps, that is, until a time of 6.25 s.

At the end of the session, we have a vector of length 200 giving the forces selected. These are the dependent variables in our policy gradients training. We also have a matrix of size 200×4 for the system states, these are our independent variables, and the vector of rewards of length 200. We sum these rewards to get the weighting for the 200 cases. This allows moves that do not maximize the immediate reward but improve the overall reward to be reinforced.

There is one more wrinkle to the training procedure. Before we begin to train the model, we want to inject some randomness into its behavior to explore the solution space. For this reason, we select a random move, no matter what the model suggests, with some probability during the early sessions of the model execution. This is done so that solution space can be explored no matter how the weights and biases in the model are initialized. As we build up more sessions, we decrease the probability of selecting a purely random force because the model will have seen more behavior that leads to good rewards to have reinforced.

Exploration

In reinforcement learning there is a necessity to explore the possible solution space to find higher rewards. This is in tension with the desire to use the model's estimate of what the best action to take is. Therefore, in training we need to allow some random moves to be made that disregard the model's

(continued)

estimate. As the model is trained, we can reduce the number of moves that are randomly selected as the reward space will be explored via random chance in the early cases seen by the model.

We ran 250 sessions of the model. We ran 50 sessions where every move made was randomly selected and then trained the model with the Adam optimizer for 20 epochs and a learning rate of 0.001. Then we ran 50 sessions where only 32% of the moves were randomly selected, and the others were chosen based on the model's output probabilities, and repeated the training for 20 epochs. The final three groups of sessions had 50 sessions each with 10%, 3.2%, and 1% probabilities of a random force applied. After each set of 50, we train the model for 20 epochs. Before each round of training, we normalize the rewards by subtracting the mean reward divided by the standard deviation of rewards.

The resulting model is able to find a policy that swings the mass to its maximum possible height. Because the moves are still selected based on the probabilities of selecting a given force, running the model different times gives different outcomes. The outcome of running the model for the cart-mounted pendulum is shown in Fig. 9.3. In the figure we show the system at several different times, depicting the force applied and using partial transparency to show several previous times to give a sense of how the system is evolving. In Fig. 9.3, at $t = 0.4$ s, the cart moves to the left, before applying a force in the opposite direction to give the mass further angular momentum (see $t = 0.9$ s panel). This oscillation between directions of applied force is repeated until eventually the mass has enough momentum to reach its maximum height around 3.4 s.

For a different view of the strategy found by the neural network via reinforcement learning, in Fig. 9.4 we show the evolution of the vertical and horizontal positions of the pendulum and the force applied by the cart. In the top panel of the figure, we can see that initially the pendulum has relatively modest swings that are successively amplified until the third period where it passes through the maximum height, before having a small amplitude path, that leads to a large swing again (this last swing is occurring when the final step in the session is reached).

The second panel of Fig. 9.4 shows that initially the cart moves to the left, before its velocity switches to the positive x-direction and back again to a negative acceleration. Then at about $t = 2$ s, the cart accelerates in the positive direction to accomplish the swing of the pendulum through the positive vertical, before switching directions again at about $t = 4.5$ s. All of these changes in direction can be seen in the third panel where the forces are displayed. In this part of Fig. 9.4, we can also see how there is still some randomness in the behavior of the system due to the model returning the probability of applying a force. Despite the randomness, there is a pattern to these forces: initially negative and then positive, before switching to negative for a short period of time, before a long period of positive force applied starting around $t = 2.5$ s. One thing we notice in the force applied is that there

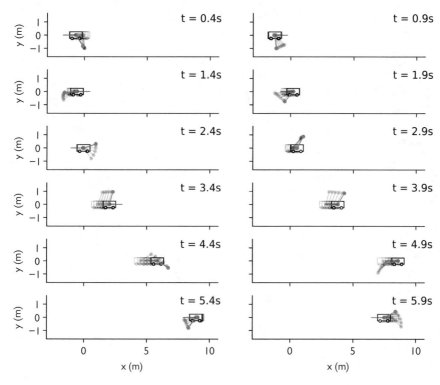

Fig. 9.3 Depiction of the strategy learned by reinforcement learning for the cart-mounted pendulum problem. The cart and the pendulum are shown at equally spaced times. The system at several previous frames is also shown as partially transparent to give a sense of how the system is changing

are relatively few times where the force applied is 0. Many of these occur during the time when the mass is near its maximum height. It is possible that the model is trying to balance the pendulum at the peak height, but that the forces that it can apply are too large for the fine control needed for this balancing.

Before moving on from the cart-mounted pendulum problem, we compare the model's strategy to the strategy from control theory. The swing-up of the pendulum from rest requires a nonlinear control strategy, and balancing it can be dealt with via linear controls [2]. This suggests that it might be beneficial to train a different model for the balancing of the pendulum and apply that when the pendulum is near vertical.

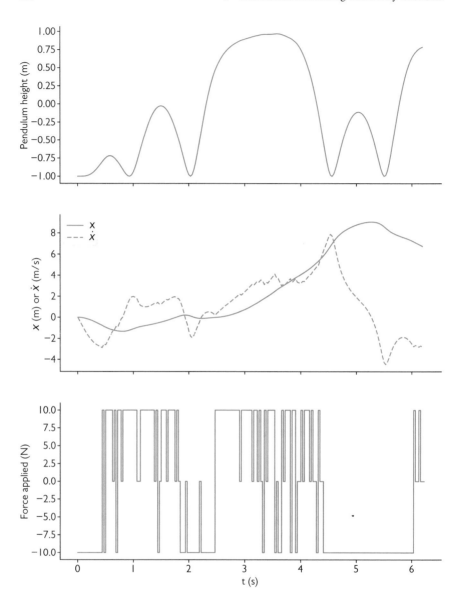

Fig. 9.4 The evolution of the pendulum height (top), cart position and velocity (middle), and the force applied (bottom) as a function of time from the strategy employed by the machine learning model

9.4   Case Study: Control of Industrial Glass Cooling with Policy Gradients

One step in the manufacturing of glass involves the cooling of the glass after it is formed. In glass different temperature levels promote different chemical reactions and physical transformations, and temperature gradients can induce stresses in the glass that lead to the development of cracks. Additionally, because the temperatures of the medium during the cooling are high, in the 1000 K range, much of the heat transfer occurs via thermal radiation [3–5]. In this case study we use reinforcement learning to develop a control strategy for cooling of glass along a specified temperature curve. Our problem is based on a problem presented in the control work of Frank et al. [5].

We use a 1D diffusion model for the radiative transfer from the hot glass where we assume that the glass is a sheet that is much larger in two dimensions and thin in a single direction, x. The equation that governs the gray energy density of the thermal radiation, $E(x, t)$ in units of J/m^3, is

$$\frac{\partial E}{\partial t} - \frac{c}{\kappa}\frac{\partial^2 E}{\partial x^2} = -c\kappa(E - aT^4), \tag{9.13}$$

where c is the speed of light with units of m/s, κ is the absorption opacity with units of m^{-1}, $T(x, t)$ is the temperature in K, and

$$a = \frac{4\sigma_{\mathrm{SB}}}{c} = 7.56438 \times 10^{-16} \ \frac{\mathrm{J}}{\mathrm{m}^3 \, \mathrm{K}^4} \tag{9.14}$$

is the radiation constant with $\sigma_{\mathrm{SB}} \approx 5.67 \times 10^{-8}$ W/(m^{2}K^4) denoting the Stefan–Boltzmann constant. We consider that the slab is symmetric about $x = 0$ and that at $x = L$ the boundary condition imposes that there is incoming radiation equivalent to that being emitted by a blackbody at temperature u. The temperature of the slab is governed by an equation that couples to Eq. (9.13):

$$\rho c_m \frac{\partial T}{\partial t} - k\frac{\partial^2 T}{\partial x^2} = c\kappa(E - aT^4), \tag{9.15}$$

with c_m the specific heat in J/kg, ρ the glass density in kg/m^3, and k the thermal conductivity in W/(m·K). As with the radiation energy density equation, the glass temperature is symmetric at $x = 0$. The boundary condition at $x = L$ is a convective boundary condition that provides

$$k\left.\frac{\partial T}{\partial x}\right|_{x=L} = h(u - T), \tag{9.16}$$

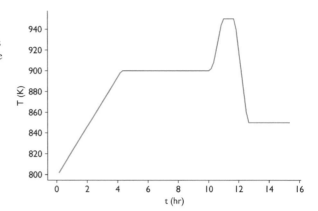

Fig. 9.5 Goal temperature for the glass sheet as a function of time for the glass cooling control problem. The temperature profile has several temperature levels held for different periods of time to promote different chemical reactions or physical changes

where h is the convection coefficient in units of W/(m$^2\cdot$K). Note that these equations are nonlinear in the temperature due to the T^4 term that accounts for radiation emission.

In this model we can control the temperature of the glass only by setting the temperature outside the boundary of the glass u. When u changes, the amount of radiation entering the glass changes and the flow of heat from convection will also change the temperature. Raising the value of u will increase the temperature of the glass, but the added heat will have to diffuse in from the boundary over time. As a result, there is a finite time for the changes of the temperature on the boundary to propagate through the medium and cause the temperature to reach a new steady state.

The values of the glass properties that we use for our case study are

- the half-thickness of the glass sheet is $L = 0.2\,\mathrm{m}$,
- the density of the glass is $\rho = 2200\,\mathrm{kg/m}^3$,
- the specific heat of the glass is $c_m = 900\,\mathrm{J/kg}$,
- the thermal conductivity is $k = 22.27\,\mathrm{W/(m\cdot K)}$,
- the convection coefficient is $h = 100\,\mathrm{W/(m}^2\cdot\mathrm{K)}$, and
- the absorption opacity is $\kappa = 100\,\mathrm{m}^{-1}$.

In the case study, we desire to have the glass temperature follow the temperature profile given in Fig. 9.5. This profile has the glass temperature start at 800 K and then increase over several hours to 900 K, before a steep temperature rise to 1000 K around 10 h, and staying there for about 1 h before rapidly cooling to 850 K. A profile such as this could be designed to control chemical reactions or physical changes that take place at different rates at 900, 1000, and 850 K. We also desire the entire volume of glass to have the same temperature, but, as we explained above, temperature changes at the boundary require a finite amount of time to propagate through the glass.

Our goal, G, is to minimize the squared difference between the glass temperature and the desired temperature integrated over the glass thickness:

$$G(t) = \int_0^L (T(x,t) - T_{\text{goal}}(t))^2 \, dx. \qquad (9.17)$$

We use this goal to define the reward function to define the reinforcement learning problem as the inverse of $G(t)$:

$$R(t) = \frac{1}{G(t) + \epsilon}, \qquad (9.18)$$

where $\epsilon = 10^{-3}$ is used to set a maximum reward for a given time.

To determine the temperature of the glass as a function of space and time and the reward function, we solve Eqs. (9.13) and (9.15) with a finite volume discretization, and a semi-implicit linearization for the nonlinear terms, and backward Euler for integrating the other terms in time. We consider that the boundary temperature is constant over an interval of 10 min. We also discretize in space with 100 zones, for a zone spacing of 0.02 m.

Our control problem is to set the value of u based on a machine learning model. For inputs to the model, we provide 8 independent variables at time t as follows:

- the current mean temperature in the slab,
- the standard deviation of the spatial variation of the temperature in the slab at a given time,
- the current value of u, and
- the target temperature at 10, 20, 30, 40, and 50 min in the future (5 variables).

The machine learning model will output 7 probabilities: the probability of changing u by ± 40, ± 20, ± 5, or 0 K. In this manner, the control can only move in a finite number of increments and can only adjust from the current state.

The model we use is a relatively simple neural network. We believe this will work because in other optimal control research on similar problems, a linear, though based on sophisticated application of adjoints, control was demonstrated to be sufficient [3, 5]. Our model has a single hidden layer with 10 units and a softmax output layer with 7 units corresponding to the 7 different possible changes to the boundary temperature. We use policy gradients to train the model, where the weighting we use is the cumulative reward for a case as the sum of the reward over 93 steps, each step having a size 10 min. That is, after each step of 10 min, we evaluate the model to get probabilities for each change to make and then sample from those probabilities to see how to change u for the next 10 min increment.

To explore the solution/reward space, we train the model in five increments. In each increment, there are 50 cases where at each step the probability of disregarding the model and using a random move to change u is 50 sessions each with 100%, 32%, 10%, 3.2%, and 1%. After each group of 50 cases, we update the model using the Adam optimizer, a learning rate of 0.001, and 20 epochs. After training the model, there is still some randomness when we use it because the model only provides probabilities for a change being correct.

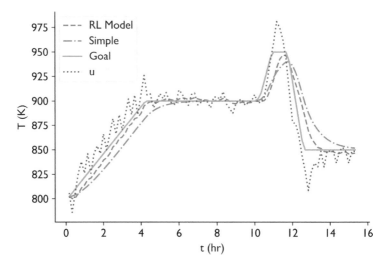

Fig. 9.6 Temperature at $x = L/2$ as a function of time produced by applying the reinforcement learning model to control the temperature, the simple control, the desired temperature, and the control temperature u selected by the reinforcement learning model

After training, we evaluate the model by running 10 cases and averaging the results to see how the model will perform on average. This serves to smooth the randomness in the application of the model. We compare the result from reinforcement learning with the intuitive, simple control of setting the value of u to be the desired temperature at time t [5]. In Fig. 9.6, we show the temperature value at $x = L/2$ over time using the reinforcement learning (RL) control of the boundary temperature and compare that to the ideal temperature and the temperature resulting from the simple control. This figure also shows the boundary temperature set by the reinforcement learning model.

From Fig. 9.6, we see that on the whole the RL control is closer to the ideal temperature than that resulting from the simple control. This is especially clear in the initial ramp up to 900 K, the peak temperature reached, and the cooling to 850 K. We can see that the model has figured out that to match the initial temperature ramp it must raise the boundary temperature above the desired temperature because the desired temperature is rising faster than the glass temperature can respond. At the plateau at 900 K, the RL control raises and lowers the boundary temperature to attempt to make the overall temperature settle near the desired temperature. We also see that there are limitations to the control system we have. Given the slow response of the glass temperature to the boundary control, the boundary temperature must rise very quickly to go up to 1000 K and decrease very quickly to go down to 850 K.

We can further investigate the system behavior by looking at the temperature as a function of space at a given time. The temperature as a function of space at time $t = 12$ h and 50 min is shown in Fig. 9.7. In this figure, we can observe that the RL control has adjusted the temperature so that it intersects the desired temperature. The

Fig. 9.7 Temperature at $t = 12$ h and 50 min as a function of space produced by applying the reinforcement learning model to control the temperature, the simple control, and the desired temperature

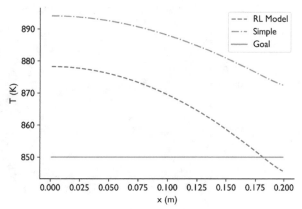

time depicted in the figure is soon after the quick cooling from 1000 K to 850 K. The strategy used by the RL control is able to get to this lower temperature by setting the boundary temperature far below the desired temperature.

A final way to judge the behavior of our control strategy is by looking at the difference between the temperatures obtained from the RL control strategy compared with the desired temperature over all time and over the entire volume. These differences are depicted in Fig. 9.8 for the RL control and the simple control. From the MAE and MSE of each temperature field, we see that the RL control is superior to the simple control. Nevertheless, one feature we notice in the results from the RL control is the rapid changes in the temperature near the boundary at $x = 0.2$ m due to the rapid changing of the boundary temperature by the RL control. Though the control makes the temperature difference close to zero, the rapid changes would likely be less than desirable for the cooling of glass in an industrial process. However, we did not specify in our reward function that the temperature should not change rapidly over time. If we did desire this property, we could change the reward and retrain the model.

In Fig. 9.8, we also see that the largest errors are at low values of x. This is due to the fact that changes in the boundary temperature take time to reach the center of the plate. Finally, we can also observe that error in the RL control temperature is larger in the rise to 1000 K between 10 and 11 h, but that it has a longer region of small error than the simple control.

Conclusions from the Case Study

This case study involved the control of a system of partial differential equations to obtain a given solution profile. The reinforcement learning model based on policy gradients had no knowledge of the underlying equations for the temperature of the glass, it only knew what it learned from changing the boundary temperature and seeing how the reward changed. In this particular problem, solving the system of equations was not too burdensome because the numerical solution was not expensive. If each trial, that is the simulation of the glass temperature over the

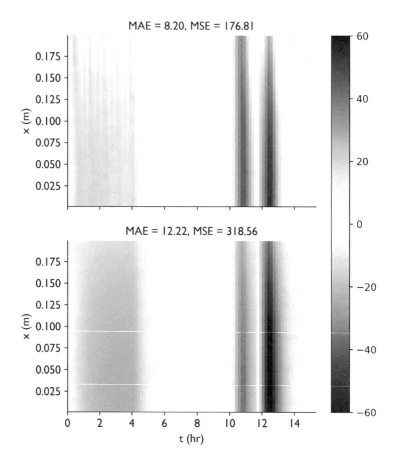

Fig. 9.8 The difference between the desired temperature and the temperatures obtained by the different controls at all times and over the entire volume for the RL control (top panel) and the simple control (bottom panel). The mean-absolute error and mean-squared error for each of the temperatures are also given

entire cooling period, required a large amount of computing resources, applying reinforcement learning may have been prohibitively expensive.

Additionally, training the model may not be the end of the story if we wanted to apply what it learned to an actual industrial process. We might reformulate the problem with larger possible changes to the boundary temperature than the 7 we used. We might also tweak the reward function to penalize unnecessary changes to boundary temperature to stop the rapid change in time we observed in the temperature near the boundary.

Notes and Further Reading

Policy gradients are only one type of reinforcement learning problem that can be applied in science and engineering. Coverage of the common approaches can be found in the text by Sutton and Barto [6]. There is a sophisticated theory behind the ideas of using randomness to explore the solution space (like we did in our cart-mounted example), the structure of the rewards, and the number of cases that must be used to train the model.

Problems

9.1 Repeat the cart-mounted pendulum problem, but start the pendulum at the top of swing with a small perturbation in θ. Use policy gradients to train a control strategy for balancing the pendulum using the procedures outlined above.

9.2 A variant on the cart-mounted pendulum problem involves a mechanical gantry that needs to move a mass from a starting position to another position in a minimum amount of time. The time required to move the mass includes the time for the mass to settle (i.e., any swinging to stop). Using policy gradients train a control strategy for a mechanical gantry where the reward is the inverse of the time required to reach the endpoint times the stopping point (to account for the fact that longer distances will require more time). The inputs to the model are the current position, the desired stopping position, and the model outputs are the probability that a force of -10, 0, or 10 N is applied. Say the mass of the gantry is 1 kg, the mass of the load is 1 kg, and the arm length is 1 m. When training the model, you will need to give it various values for the desired stopping position. Can your model accomplish the goal?

References

1. Volodymyr Mnih, Koray Kavukcuoglu, David Silver, Andrei A Rusu, Joel Veness, Marc G Bellemare, Alex Graves, Martin Riedmiller, Andreas K Fidjeland, Georg Ostrovski, et al. Human-level control through deep reinforcement learning. *Nature*, 518(7540):529, 2015.
2. Russ Tedrake. Underactuated robotics: Learning, planning, and control for efficient and agile machines course notes for MIT 6.832. 2009.
3. Guido Thömmes, René Pinnau, Mohammed Seaid, Th. Götz, and Axel Klar. Numerical methods and optimal control for glass cooling processes. *Transport theory and statistical physics*, 31(4-6):513–529, 2002.
4. Axel Klar, Jens Lang, and Mohammed Seaïd. Adaptive solutions of -approximations to radiative heat transfer in glass. *International Journal of Thermal Sciences*, 44(11):1013–1023, November 2005.
5. Martin Frank, Axel Klar, and René Pinnau. Optimal Control of Glass Cooling Using Simplified PN Theory. *Transport Theory and Statistical Physics*, 39(2-4):282–311, March 2010.
6. Richard S Sutton and Andrew G Barto. *Reinforcement learning: An introduction*. MIT press, 2018.

Appendix A
Data and Implementation of the Examples and Case Studies

The data and code required to reproduce the examples and case studies in this monograph are available at the author's page on Github: https://github.com/DrRyanMc. The files there are arranged chapterwise. Where possible, random number seeds have been set to make reproducible results, but this has not always been possible. The codes are in the form of Jupyter notebooks, written in Python. The examples below assume a basic knowledge of Python.

A.1 Scikit-Learn

Many of the examples from Part I were produced using the tools available in Scikit-learn. Scikit-learn is a library of machine learning methods and algorithms that make the use of machine learning in Python rather straightforward. This library can be found at https://scikit-learn.org/. Below is some discussion of the features and use of Scikit-learn. It is not meant to be exhaustive, nor is it a replacement for the fine and thorough documentation available in the Scikit-learn website.

A.1.1 Training and Using ML Models

Scikit-learn is structured with a common syntax for the different models that it employs (e.g., linear regression, logistic regression, random forests, etc.) in that there is a common way to initialize, train, and use the models. For instance, the Python code to initialize and train a linear regression model with training data X and y, for the independent and dependent variables, would be

© Springer Nature Switzerland AG 2021
R. G. McClarren, *Machine Learning for Engineers*,
https://doi.org/10.1007/978-3-030-70388-2

```
1  from sklearn.linear_model import LinearRegression
2  reg = LinearRegression()
3  reg.fit(X, y)
```

In this code snippet, line 1 imports the linear regression function, line 2 defines `reg` as a linear regression model using the default parameters, and the third line trains the model with independent variables contained in X and dependent variables in y.

Performing the same procedure with random forests looks almost identical:

```
1  from sklearn.ensemble import RandomForestRegressor
2  rf = RandomForestRegressor()
3  rf.fit(X, y)
```

In this case, `rf` is a random forest model that we train by invoking the `fit` method.

Similarly, we can make a prediction using either of these models using the `predict` method:

```
1  ytestReg = reg.predict(Xtest)
2  ytestRF = rf.predict(Xtest)
```

where we have assumed that `Xtest` is a matrix of test data that we want to use to evaluate the model.

The beauty of this standardized form is that we can easily swap out different models with only a small change of syntax. We can even perform unsupervised learning using the same structure.

To apply K-means to a data set, the syntax is similar to a supervised learning model, except we do not supply dependent variables:

```
1  from sklearn.cluster import KMeans
2  kmeans = KMeans(n_clusters=2)
3  kmeans.fit(X)
4  clusters = kmeans.predict(Xtest)
```

In this code we apply K-means with $K = 2$ to the data in X. Notice that in line 3 we use the same `fit` method but only pass a single variable. We then predict the cluster membership for some test data and store the cluster IDs in `clusters`. The same basic syntax and structure is used for other unsupervised learning techniques such as t-SNE.

All of the models discussed above have parameters that can be set when the method is initialized. For instance, a random forest model will have parameters for the number of trees in the forest. The syntax for setting these parameters can be found in the Scikit-learn documentation for the particular model.

A.1.2   Evaluating Model Performance and Cross-Validation

In Scikit-learn, there are functions that automate some common model assessment tasks. For example, one can perform K-fold cross-validation using a single function:

```
1  from sklearn.model_selection import cross_validate
2  result = cross_validate(reg, X, y, cv=3)
```

Here we take the linear regression model `reg` defined above and perform 3-fold cross-validation on the model: It will divide the data into three subsets, train the model on each, and report the loss function in the variable `result`. This type of cross-validation could be used for any of the models we discussed previously in this section.

One important use of cross-validation is to select the penalty parameter in regularized regression. This can be done automatically in Scikit-Learn. Here we apply 10-fold cross-validation to a lasso model to pick this parameter in a single training step:

```
1  from sklearn.linear_model import LassoCV
2  lasso = LassoCV(cv=10)
3  lasso.fit(X, y)
4  yhat = lasso.predict(X)
```

Using the `LassoCV` method, the `fit` method performs the cross-validation, picks the penalty parameter, and returns the best model. Then we can use this model with the `predict` method, as before.

There are additional capabilities to perform cross-validation on any parameter in a Scikit-learn model using the function `RandomizedSearchCV`. This function takes as arguments the model that cross-validation should be performed on, and the parameters that should be varied in the cross-validation. The result is the best model found by cross-validation. See the Scikit-learn documentation for the full list of features for this function.

A.1.3   Preprocessing and Loading Data

Another benefit of Scikit-learn is that many data preprocessing tasks can be handled with pre-defined routines. Scaling, normalizing, and other transformations to data can be accomplished by the `Preprocessor` module in Scikit-learn. Additionally, the partitioning of data into training and test sets can be done using a built in function:

```
1  from sklearn.model_selection import train_test_split
2  Xtrain,Xtest,ytrain,ytest = train_test_split(X, y,
3                                              test_size=0.2)
```

In this code snippet, the independent variables in X and the dependent variables in y are split into test and training sets where 20% of the data is randomly assigned to the test set.

Finally, the `datasets` module in Scikit-learn contains a variety of standard datasets used in machine learning problems, such as the handwritten images data set, and various other image, classification, or regression data sets. Additionally, there are functions that can load data directly from existing ML data repositories.

A.2   Tensorflow

Tensorflow is a library for building deep neural networks (and related models) in Python. It was originally developed by Google, but released in 2015 under a public use license. The website for Tensorflow is https://www.tensorflow.org. The majority of the neural networks used in this work can be constructed with the `keras` module of Tensorflow. Keras is another open-source machine learning library, and Tensorflow has its own implementation of the library. We refer to the Keras module in Tensorflow as `tf.keras`.

In `tf.keras` we can construct a deep neural network sequentially. For instance, we can define a feed-forward, fully connected neural network with 4 inputs, 2 outputs, and 3 hidden layers each with 6 hidden units and the ReLU activation function as

```
1  import tensorflow as tf
2  model = tf.keras.Sequential()
3  model.add(tf.keras.layers.Dense(6, activation='relu',
4            input_shape=(4,)))
5  model.add(tf.keras.layers.Dense(6, activation='relu'))
6  model.add(tf.keras.layers.Dense(6, activation='relu'))
7  model.add(tf.keras.layers.Dense(2))
```

Line 2 of this code defines `model` as a `tf.keras` model that will be built sequentially. The first item added to this model is a fully connected, or "dense," hidden layer of 6 nodes with the ReLU activation function that takes an input of size 4, as defined in lines 3 and 4. Lines 5 and 6 add two subsequent hidden layers, before the output layer of size 2 (i.e., with two outputs) finalizes the model in line 7. If an activation function is not specified, then the identity activation is used.

To train the model we first compile it by giving it training parameters, and then pass it training data to train:

```
1  model.compile(loss='MSE',
2                optimizer='adam',
3                metrics=['MAE'])
4  model.fit(x=X, y=Y, epochs=20, batch_size=32)
```

When we compile the model we specify the loss function, here we use the mean-squared error loss function, and specify the optimization method, Adam with the default parameters in this case. We can also specify other metrics that we want to track during training. Here we keep track of the mean-absolute error as well. Line 4 in the above performs the training by calling the `fit` method. There we specify the inputs as `x` and the dependent variables for training as `y`. For these parameters we specify that the independent variables are in a variable called X and the dependent variables are in Y. Additionally, we specify how many epochs to train for and the batch size.

When we have a trained model, we can use it to predict new dependent variables using the `predict` method:

```
1  Yhat = model.predict(X)
```

Convolutional neural networks can be built in almost identical fashion, but with different layer types. The model used to predict the class for the MNIST fashion data set in Chap. 6 is defined by

```
1   model = tf.keras.Sequential()
2   model.add(tf.keras.layers.Conv2D(filters=64, kernel_size=2,
3            padding='same', activation='relu',
4            input_shape=(28,28,1)))
5   model.add(tf.keras.layers.MaxPooling2D(pool_size=2))
6   model.add(tf.keras.layers.Dropout(0.3))
7   model.add(tf.keras.layers.Conv2D(filters=32, kernel_size=2,
8            padding='same', activation='relu'))
9   model.add(tf.keras.layers.MaxPooling2D(pool_size=2))
10  model.add(tf.keras.layers.Dropout(0.3))
11  model.add(tf.keras.layers.Flatten())
12  model.add(tf.keras.layers.Dense(256, activation='relu'))
13  model.add(tf.keras.layers.Dropout(0.5))
14  model.add(tf.keras.layers.Dense(10, activation='softmax'))
```

As before, this model is built sequentially. The first layer we add is a 2-D convolution of kernel size 2 that produces 64 channels (this is the number of filters in `tf.keras`), padded to keep the same size, using the ReLU activation function, and expecting an input of size 28×28. We then add max pooling and dropout layers. The `pool_size` parameter will shrink the image by a factor of 2 in each dimension (down to 14×14), and the dropout layer applies dropout with a probability of 30% to the connections from the max pooling layer to the next layer, a 2-D convolution producing 32 channels of size 14×14 each.

Eventually, the data is flattened (in line 11) into a vector that is fed into a fully-connect layer containing 256 hidden units (in line 12). This layer then has dropout applied to it with probability of 50% of dropping a connection, before being passed to an output layer with a softmax activation function to produce a set of probabilities for each of the 10 classes. To train this model, one would compile and fit as done for the fully connected example above.

There are many possible layers that can be included in a `tf.keras` model. These include recurrent units, long short-term memory units, transposed convolutions, and all the other types of units discussed in the text. The Tensorflow documentation gives descriptions of how to use each of these.

It is possible to construct models without making them sequentially. The following code constructs a neural network with 50 independent variables as inputs and 4 outputs by defining intermediate steps:

```
1   inpts = tf.keras.Input(shape=(50,1), name='inputs')
2   x = tf.keras.layers.Conv1D(2, kernel_size=2)(inputs)
```

```
3   x1 = tf.keras.layers.Conv1D(4,kernel_size=2,padding="same")(x)
4   x2 = tf.keras.layers.Conv1D(8,kernel_size=2,padding="same")(x1)
5   x3 = tf.keras.layers.Flatten()(x2)
6   x4 = tf.keras.layers.Dropout(rate=0.05)(x3)
7   outpts = tf.keras.layers.Dense(4, activation="softplus")(x4)
8   model = tf.keras.Model(inputs = inpts, outputs = outpts)
```

In this code inpts is defined as the input to a model in line 1. Then a succession of variables, x, x1, x2, x3, and x4 are defined as the outputs from hidden layers. Finally, we define a variable called outpts that outputs 4 numbers passed through the softplus activation function. In line 8 the model is constructed by specifying the inputs and outputs to the tf.keras model. This model can then be trained and used to make predictions as discussed above.

With these simple instructions, we can build a variety of sophisticated deep learning models. Of course, this discussion only scratches the surface of the capabilities in Tensorflow. Autoencoders, transfer learning, and reinforcement learning are all possible using slight modifications to the models defined above. See the code for the examples provided via Github for this book, or the Tensorflow documentation for more information.

Index